Lecture Notes in Mathematics

A collection of informal reports and seminars
Edited by A. Dold, Heidelberg and B. Eckmann, Zürich

Series: Department of Mathematics, University of Maryland, College Park
Adviser: J. K. Goldhaber

200

Umberto Neri

University of Maryland, College Park, MD/USA

Singular Integrals

Springer-Verlag
Berlin · Heidelberg · New York 1971

AMS Subject Classifications (1970): 46E35, 47G05

ISBN 3-540-05502-9 Springer-Verlag Berlin · Heidelberg · New York
ISBN 0-387-05502-9 Springer-Verlag New York · Heidelberg · Berlin

Offsetdruck: Julius Beltz, Hemsbach/Bergstr.

PREFACE

Art, said Plato, is a copy of a copy. So are these notes. They form the content of a course I gave this semester at the University of Maryland, but the material is derived from courses held by Professor Antoni Zygmund at the University of Chicago (1963-64) and at the University of Paris (Orsay, 1965).

Because any introduction should be elementary, the only background needed by the reader is a knowledge of Lebesgue Integration and some acquaintance with Complex Variable and the simplest concepts of Functional Analysis.

I wish to thank Professor Robert Freeman who proofread the entire manuscript and Mrs. Joseph Lehner who typed it.

U. Neri

College Park, May 1967

CONTENTS

PART I. SINGULAR INTEGRALS: AN INTRODUCTION

PART II. SINGULAR INTEGRAL OPERATORS AND DISTRIBUTIONS

PART I

SINGULAR INTEGRALS: AN INTRODUCTION

INTRODUCTION

The subject of Singular Integrals has undergone a considerable developement in the last twenty years, as a result of the works of Mihlin, Calderòn and Zygmund. It has shown to have important connections with the study of Fourier Series, Partial Differential Equations, and other branches of Analysis.

A classical example of a singular integral is given by the Hilbert transform

$$\tilde{f}(x) = \frac{1}{\pi} \int_{-\infty}^{+\infty} \frac{f(t)}{x - t}\, dt \qquad -\infty < x < +\infty$$

Even if f is a "very good" function, the integral, as it stands, does not exist, due to the singularity of the integrand at $x = t$. However, following Cauchy, we may consider this integral in the "principal value sense" (p.v.), by removing a symmetric neighborhood of the singularity; namely,

$$\tilde{f}(x) = \text{p.v.}\ \frac{1}{\pi} \int_{-\infty}^{+\infty} \frac{f(t)}{x - t}\, dt = \lim_{\varepsilon \to 0^+} \frac{1}{\pi} \int_{|x-t| > \varepsilon} \frac{f(t)}{x-t}\, dt.$$

This singular integral (or principal value integral) arose in the study of boundary values of analytic functions. We have two kinds of problems here.

I. Existence of $\tilde{f}(x)$.

II. Relationship between f and $\tilde{f}$: that is, what properties of f are kept by $\tilde{f}$?

PRELIMINARIES

Let $x = (x_1, \ldots, x_n) \in E^n$, where E^n is Euclidean n-dimensional space, with Lebesgue measure $dx = dx_1 \ldots dx_n$. All sets and functions considered will be Lebesgue measurable. If $f(x)$ is a measurable function defined a.e. (almost everywhere) on a set $S \subset E^n$, we consider the integral

$$\int_S f(x)\,dx = \int_S f(x_1, \ldots, x_n)\,dx_1 \ldots dx_n.$$

If we set $f(x) = 0$ outside S, we may write the integral as

$$\int_{E^n} f(x)\,dx = \int f(x)\,dx$$

where the domain of integration is understood to be the entire space E^n.

Letting $|x| = (\sum_1^n x_i^2)^{1/2}$ denote the norm of x, consider the integrals

$$\int \frac{dx}{|x|^\alpha}, \quad \alpha > 0.$$

The integrands may have trouble at 0 and at infinity, so we split the integral into two parts:

$$I_1 = \int_{|x| \leq 1} \frac{dx}{|x|^\alpha} \text{ and } I_2 = \int_{|x| \geq 1} \frac{dx}{|x|^\alpha}.$$

Letting $r = |x|$, the element of volume dx can be written in "polar coordinates" as $dx = r^{n-1} d\sigma dr$, where $d\sigma$ is the surface element on the unit sphere $\Sigma = \{x : |x| = 1\}$. Thus,

$$I_1 = \int_{|x| \leq 1} \frac{dx}{|x|^\alpha} = \int_\Sigma d\sigma \int_0^1 \frac{r^{n-1} dr}{r^\alpha} = C \int_0^1 r^{n-\alpha-1} dr < \infty \quad \text{if}$$

and only if $n - \alpha > 0$, that is $\alpha < n$. (Here the constant C is the area of the unit sphere).

On the other hand,

$$I_2 = \int_{|x| \geq 1} \frac{dx}{|x|^\alpha} = \int_\Sigma d\sigma \int_1^\infty r^{n-\alpha-1} dr = C \int_1^\infty r^{n-\alpha-1} dr < \infty \quad \text{if}$$

and only if $n - \alpha < 0$, that is $\alpha > n$.

If $\alpha = n$, then $I_1 = I_2 = \infty$.

With $1 \leq p < \infty$, we consider the Banach spaces $L^p = L^p(E^n) = \{f: \int |f(x)|^p dx < \infty\}$, wher the functions f are measurable and complex-valued (in general), and two functions are identified if they coincide a.e. . The norm

$\|f\|_p = \{\int |f|^p dx\}^{1/p}$ satisfies the

<u>Triangle Inequality</u>: $\|f + g\|_p \leq \|f\|_p + \|g\|_p$ (Minkowski). Also, we have

<u>Hölder's Inequality</u>: $|\int fg \, dx| \leq \|f\|_p \|g\|_{p'}$, where $p > 1$ and p', the conjugate exponent of p, is defined by the relation

$$\frac{1}{p} + \frac{1}{p'} = 1.$$

If $p = 1$, then Hölder's inequality holds with $p' = \infty$, where we set

$\|g\|_\infty = \text{ess sup } |g(x)|$ = the least number $M \geq 0$ such that $|g(x)| > M$ only on a set of measure zero,

and define the Banach space L^∞ accordingly. We recall that the spaces L^p and $L^{p'}$ are dual, with the dual pairing

$$\langle f,g \rangle = \int f(x) \, g(x) \, dx.$$

If $E \subset E^n$, one defines similarly the spaces $L^p(E)$; for these spaces the above facts also hold. Let $|E|$ denote the measure of E. If $|E| < \infty$, we can modify the definition of norm in $L^p(E)$ by considering the average

$$A_p[f] = \left(\frac{1}{|E|}\int_E |f|^p\right)^{1/p}.$$

THEOREM: Let $E \subset E^n$, $|E| < \infty$. If $r < s$ then $L^s(E) \subset L^r(E)$ and

$$A_r[f] \leqslant A_s[f].$$

Proof: Let $p = \frac{s}{r} > 1$, and q be such that $\frac{1}{p} + \frac{1}{q} = 1$. By Hölder's inequality

$$\int_E |f|^r dx = \int_E |f|^r \cdot 1 dx \leqslant \left(\int_E |f|^{rp} dx\right)^{1/p}\left(\int_E 1^q dx\right)^{1/q}$$

$$= |E|^{1/q}\left(\int_E |f|^{rp}\, dx\right)^{1/p}.$$

Since $q = p/(p-1) = s/(s-r)$ and $rp = s$, we have

(1) $$\int_E |f|^r\, dx \leqslant |E|^{\frac{s-r}{s}}\left(\int_E |f|^s\, dx\right)^{\frac{r}{s}}$$ which implies that $L^s(E) \subset L^r(E)$.

Moreover, from (1) we obtain

$$\frac{1}{|E|}\int_E |f|^r\, dx \leqslant \frac{1}{E^{r/s}}\left(\int_E |f|^s\right)^{r/s}$$, whence taking r^{th} roots,

$$A_r[f] \leqslant A_s[f]$$

Q.E.D.

REMARK: $A_p[f] \to \operatorname*{ess\,sup}_{x\in E} |f(x)| = \|f\|_\infty$ as $p \to \infty$.

If $|E| = \infty$, the $L^p(E)$ spaces are no more comparable. For instance $f(x) = 1/x$ belongs to $L^2(1,\infty)$ but does not belong to $L^1(1,\infty)$.

From Hölder's inequality it follows by induction that if p_i are positive numbers with $\frac{1}{p_1} + \frac{1}{p_2} + \ldots + \frac{1}{p_n} = 1$, and if $f_i \in L^{p_i}$, then

$$\left|\int f_1 f_2 \ldots f_n \, dx\right| \leq \|f_1\|_{p_1} \cdot \|f_2\|_{p_2} \ldots \|f_n\|_{p_n} .$$

Another useful result is the so-called Converse of Hölder's Inequality. If $p > 1$ and $\frac{1}{p} + \frac{1}{p'} = 1$, then

$$\left|\int fg \, dx\right| \leq \|f\|_p \quad \text{for all } g \text{ with } \|g\|_{p'} = 1.$$

But if we take the supremum over all such g we attain equality:

(Converse of Hölder) $$\sup_{\substack{g \\ \|g\|_{p'}=1}} \left|\int fg dx\right| = \|f\|_p .$$

Remark: If $\mu(x)$ is another measure on E^n, we can consider integrals

$$\int f(x)\mu(dx) = \int f(x) d\mu(x)$$

and we still have the above mentioned facts for the spaces $L^p(d\mu)$.

CHAPTER I

CONVOLUTIONS

Let $f(x)$ and $g(x)$ be measurable functions of $x \in E^n$. We may assume, for simplicity, that f and g are real valued.

DEFINITION: The convolution $h = f*g$ is defined by

$$h(x) = (f*g)(x) = \int f(y)g(x-y)\,dy$$

where integration is over the whole space and $\int$ is a Lebesgue integral.

Convolution is a kind of multiplication. L^1 is not closed under pointwise multiplication; for instance, $f(x) = g(x) = (x)^{-1/2}$ belongs to $L^1(0,1)$ but $f(x)g(x) = 1/x$ does not. However L^1 is closed under the "convolution product", as the following theorem shows.

THEOREM 1: If $f, g \in L^1$, then $h = f*g$ exists a.e. and belongs to L^1 Also,

$$\|h\|_1 \leq \|f\|_1 \|g\|_1.$$

Proof:

$|h(x)| \leq \int |f(y)|\,|g(x-y)|\,dy$. Integrating in x and interchanging the order of integration (Tonelli's Theorem), since $g(x-y)$ is a measurable function of (x,y),

$$\int |h(x)|\,dx \leq \int dx\left(\int |f(y)|\,|g(x-y)|\,dy\right) = \int |f(y)|\,dy\left(\int |g(x-y)|\,dx\right) =$$

$$= \|f\|_1 \|g\|_1 < \infty$$

Hence $h(x) < \infty$ a.e. and $\|h\|_1 \leq \|f\|_1 \|g\|_1$.

Q.E.D.

Convolution is commutative, that is $f*g = g*f$. In fact,

$$(f*g)(x) = \int f(y)g(x-y)\,dy = \text{changing variable } (z = x-y,\ dz=dy)$$

$$= \int f(x-z)g(z)dz) = (g*f)(x).$$

EXERCISE: Show that, in L^1, convolution is associative, that is $(f*g)*h = f*(g*h)$. (Hint: may assume f,g,h all ≥ 0; interchange the order of integration and use an appropriate change of variable.)

REMARK: There is no function g such that $f*g = f$. However, the Dirac measure δ (which assigns mass $+\infty$ at the origin and mass 0 elsewhere) satisfies $f*\delta = f$, in the sense of distributions, so δ is an "identity".

THEOREM 2 (W.H. Young): Let $1 \leq p \leq \infty$. If $f \in L^p$ and $g \in L$, then $h = f*g$ exists a.e. and belongs to L^p; in fact

$$\|h\|_p \leq \|f\|_p \|g\|_1.$$

Proof: The case $p = 1$ was proved in Th. 1. If $p = \infty$, let $M = \operatorname{ess\,sup} |f(x)| = \|f\|_\infty$; then, for almost every x,

$$|h(x)| \leq \int |f(y)|\ |g(x-y)|dy \leq M \int |g(x-y)|dy = \|f\|_\infty \|g\|_1$$

which implies the conclusion.

Let $1 < p < \infty$ and p' be such that $\frac{1}{p} + \frac{1}{p'} = 1$. Using Hölder's inequality we have

$$|h(x)| \leq \int |f(y)|\ |g(x-y)|dy = \int |f(y)|\ |g(x-y)|^{\frac{1}{p}} |g(x-y)|^{\frac{1}{p'}} dy \leq$$

$$\leq \left(\int |f(y)|^p |g(x-y)|dy \right)^{1/p} \left(\int |g(x-y)|dy \right)^{1/p'}.$$ Hence,

$$|h(x)|^p \leq \|g\|_1^{p/p'} \left(\int |f(y)|^p |g(x-y)|dy \right).$$ Integrating in x, and then interchanging the order of integration, we obtain

$$\int |h(x)|^p dx \leq \|g\|_1^{p/p'} \int dx \left(\int |f(y)|^p |g(x-y)|dy \right) =$$

$$= \|g\|_1^{p/p'} \int |f(y)|^p dy \int |g(x-y)|\, dx. \text{ Therefore,}$$

$$\|h\|_p^p \leqslant \|g\|_1^{p/p'} \|f\|_p^p \|g\|_1 = \left(\text{ since } \frac{p}{p'} + 1 = p\left(\frac{1}{p'} + \frac{1}{p}\right) = p\right)$$

$$= \|g\|_1^p \|f\|_p^p.$$

Taking p^{th} roots the conclusion follows.

Q.E.D.

More generally, we have the following result.

THEOREM 3 (Young): Suppose that $f \in L^p$ and $g \in L^q$, where $\frac{1}{p} + \frac{1}{q} \geqslant 1$, and set $\frac{1}{r} = \frac{1}{p} + \frac{1}{q} - 1$. If $h = f*g$, then $h \in L^r$ and

$$\|h\|_r \leqslant \|f\|_p \|g\|_q.$$

Proof: If $q = 1$, then $r = p$, so we are back in Th. 2.
If $\frac{1}{p} + \frac{1}{q} = 1$, then $r = \infty$ and the conclusion follows readily from Hölder's inequality.

Consider the case when $\frac{1}{p} + \frac{1}{q} > 1$, so $1 < r < \infty$. We may assume that f and g are real valued and moreover that they are nonnegative. Thus

$$h(x) = \int f(y)g(x-y)dy \geqslant 0.$$

Let λ, μ, ν be positive numbers such that $\frac{1}{\lambda} + \frac{1}{\mu} + \frac{1}{\nu} = 1$. Write

$$h(x) = \int f(y)^{p\left(\frac{1}{p} - \frac{1}{\mu}\right)} g(x-y)^{q\left(\frac{1}{q} - \frac{1}{\nu}\right)} f(y)^{p/\mu} g(x-y)^{q/\nu} dy.$$

By Hölder's inequality with exponents λ, μ and ν, we obtain

$$h(x) \leqslant \left(\int f(y)^{\lambda p\left(\frac{1}{p} - \frac{1}{\mu}\right)} g(x-y)^{\lambda q\left(\frac{1}{q} - \frac{1}{\nu}\right)} dy\right)^{1/\lambda} \left(\int f(y)^p dy\right)^{1/\mu} \left(\int g(x-y)^q dy\right)^{\frac{1}{\nu}}$$

so

$$(*) \quad h(x) \leqslant \|f\|_p^{p/\mu} \|g\|_q^{q/\nu} \left(\int f(y)^{\lambda p\left(\frac{1}{p} - \frac{1}{\mu}\right)} g(x-y)^{\lambda q\left(\frac{1}{q} - \frac{1}{\nu}\right)} dy\right)^{1/\lambda}.$$

Since we want to have f^p and g^q in the integrand above, we note that we can choose λ, μ, ν in such a way that

$$\frac{1}{\lambda} = \left(\frac{1}{p} - \frac{1}{\mu}\right), \quad \frac{1}{\lambda} = \left(\frac{1}{q} - \frac{1}{\nu}\right) \quad \text{and} \quad \frac{1}{\lambda} = \frac{1}{r}; \text{ that is}$$

(1) $\frac{1}{p} = \frac{1}{\mu} + \frac{1}{\lambda}$

(2) $\frac{1}{q} = \frac{1}{\nu} + \frac{1}{\lambda}$

(3) $\frac{1}{r} = \frac{1}{\lambda}$

In fact, adding (1) and (2) and substracting (3), we obtain $\frac{1}{p} + \frac{1}{q} - \frac{1}{r} = \frac{1}{\mu} + \frac{1}{\nu} + \frac{1}{\lambda}$, and the left side of this equality is equal to 1, by definition of r.

With these choices of λ, μ, ν, we can rewrite expression (*)as

$$|h(x)| \leq \|f\|_p^{p/\mu} \|g\|_q^{q/\nu} \left(\int f(y)^p g(x-y)^q dy\right)^{1/r}$$

Taking r^{th} powers and integrating in x,

$$\int |h(x)|^r dx \leq \|f\|_p^{(pr/\mu)} \|g\|_q^{(qr/\nu)} \int dx \int f(y)^p \, g(x-y)^q \, dy,$$

and, since f^p and g^q are integrable and nonnegative, we can interchange the order of integration obtaining

$$\int |h(x)|^r \, dx \leq \|f\|_p^{(pr/\mu)} \|g\|_q^{(qr/\nu)} \int f(y)^p \, dy \int g(x-y)^q \, dx =$$

$$= \|f\|_p^{(pr/\mu)} \|g\|_q^{(qr/\nu)} \|f\|_p^p \|g\|_q^q .$$

Taking r^{th} roots and recalling that $r = \lambda$, we have

$$\|h\|_r \leq \|f\|_p^{p/\mu + p/r} \|g\|_q^{q/\nu + q/r} =$$

$$= \|f\|_p^{p\left(\frac{1}{\mu} + \frac{1}{\lambda}\right)} \|g\|_q^{q(1/\nu + 1/\lambda)} =$$

$$= \|f\|_p \|g\|_q \text{ by (1) and (2).}$$

Hence h belongs to L^r and

$$\|h\|_r \leq \|f\|_p \|g\|_q .$$

Q.E.D.

Let $f(x)$ by a measurable function of $x \in E^n$; then, for any vector $u \in E^n$, the function $f(x + u)$ is a "translation" of $f(x)$ by the vector $-u$. The following Lemma shows that this operation of translating functions is continuous in the metric of the spaces L^p, $1 \leq p < \infty$.

LEMMA 1: If $f \in L^p(E^n)$, $1 \leq p < \infty$ then

(*) $$\|f(x + u) - f(x)\|_p = \left(\int |f(x + u) - f(x)|^p \, dx\right)^{1/p} \to 0$$

as $|u| \to 0$.

Proof: We prove the Lemma in a series of steps.

(i) If $f = g + h$ where g satisfies (*) and h has arbitrary small norm, then f satisfies (*). In fact if $\varepsilon > 0$ and $\|h\|_p < \varepsilon$, then

$$\begin{aligned} \|f(x+u) - f(x)\|_p &= \|g(x+u) - g(x) + h(x+u) - h(x)\|_p \leq \\ &\leq \|g(x+u) - g(x)\|_p + \|h(x+u)\|_p + \|h(x)\|_p < \\ &< \|g(x+u) - g(x)\|_p + 2\varepsilon < 3\varepsilon \end{aligned}$$

if $|u|$ is sufficiently small.

(ii) The Lemma holds if f is continuous and has bounded support, since if S is the support of f then

$$\|f(x+u) - f(x)\|_p \leq C \max_{x \in S} |f(x+u) - f(x)| \to 0 \text{ as } |u| \to 0.$$

(iii) The Lemma holds if f is bounded and has bounded support, since any such function is the limit of functions in (ii), so using (i) the conclusion follows.

(iv) If $f \in L^p$, we consider $f_n(x) = \begin{cases} f(x) \text{ if } |x| \leq n \text{ and } |f(x)| \leq n \\ 0 \text{ otherwise.} \end{cases}$

Then $f_n \to f$ in L^p and the conclusion follows from (i) and (iii).

Q.E.D.

EXERCISE: Show that Lemma 1 is false if $p = \infty$.

Consider a function $K_\varepsilon(x)$, depending on the parameter $\varepsilon > 0$, and the convolutions

$$(f*K_\varepsilon)(x) = \int f(y)K_\varepsilon(x-y)dy.$$

Here $K_\varepsilon(x)$ is called a Kernel.

When does $f*K_\varepsilon \to f$ as $\varepsilon \to 0$?

We shall assume that our Kernel $K_\varepsilon(x)$ satisfies conditions

a) $\int |K_\varepsilon(x)|dx \leq A$, for all ε (A is constant, independent of ε)

b) $\int K_\varepsilon(x)\, dx = 1$, for all ε

c) For any fixed $\delta > 0$, $\int_{|x|\geq\delta} |K_\varepsilon(x)|dx \to 0$ as $\varepsilon \to 0$.

Before we illustrate the meaning of these conditions we shall show that such Kernels $K_\varepsilon(x)$ are easily manufactured.

LEMMA 2. Suppose that $K(x) \in L(E^n)$ is normalized so that $\int K(x)dx = 1$. Let

$$K_\varepsilon(x) = \frac{1}{\varepsilon^n} K\left(\frac{x}{\varepsilon}\right)$$

then $K_\varepsilon(x)$ satisfies properties a), b), c).

PROOF:

$$\int |K_\varepsilon(x)|dx = \int \frac{1}{\varepsilon^n} \left|K\left(\frac{x}{\varepsilon}\right)\right| dx = \left(\text{ if } y = \frac{x}{\varepsilon} \text{ so } dy = \frac{dx}{\varepsilon^n}\right)$$

$$= \int |K(y)|dy = A \text{ for all } \varepsilon.$$

Likewise,

$$\int K_\varepsilon(x)\, dx = \int K(y)\, dy = 1.$$

So a) and b) hold.

For any fixed $\delta > 0$

$$\int_{|x|\geq\delta} |K_\varepsilon(x)|\,dx = \int_{|x|\geq\delta} \frac{1}{\varepsilon^n}\,|K(\tfrac{x}{\varepsilon})|\,dx = \int_{|y|\geq\delta/\varepsilon} |K(y)|\,dy \to 0$$

as $\varepsilon \to 0$, since $\delta/\varepsilon \to \infty$ and hence we integrate outside an expanding sphere.

So c) also holds.

Q.E.D.

EXAMPLE. With $x \in E^n$, let

$$K(x) = \frac{1}{\gamma}\,\frac{1}{(1+|x|^2)^{(n+1)/2}}$$

$K(x)$ is continuous on E^n, and $K(x) = O(|x|^{-(n+1)})$ at infinity. Hence $K(x)$ is integrable. Choose γ so that

$$\int K(x)\,dx = 1.$$

Letting

$$K_\varepsilon(x) = \varepsilon^{-n}K(\tfrac{x}{\varepsilon}), \text{ we have that}$$

$$K_\varepsilon(x) = \frac{1}{\gamma}\,\frac{\varepsilon}{(\varepsilon^2+|x|^2)^{(n+1)/2}},$$

which is called the Poisson Kernel. By the previous Lemma, it satisfies properties a), b) and c).

REMARK: If $K(x) \geq 0$ is a "bell-shaped" function on E^1, with support in $|x| \leq 1$, and such that $\int K(x)\,dx = 1$, then $K_\varepsilon(x)$ has support in $|x| \leq \varepsilon$, and since

$$\int K(x)\,dx = \int K_\varepsilon(x)\,dx = 1 \text{ for all } \varepsilon,$$

then, as $\varepsilon \to 0$, the functions $K_\varepsilon(x)$ are a sequence of "peaks" which approximate the Dirac measure δ.

THEOREM 4: Suppose that $f(x)$ is bounded and that $K_\varepsilon(x)$ satisfies properties a), b) and c). Consider the convolution

$$f_\varepsilon = f*K_\varepsilon.$$

Then, $f_\varepsilon(x) \to f(x)$, as $\varepsilon \to 0$, at every point of continuity of f and the convergence is uniform for x belonging to any compact set of points of continuity of f.

Proof: Let x be a point of continuity of f, and let $\eta > 0$ be arbitrary. For some $\delta > 0$

(1) $$|f(x-y) - f(x)| < \eta \quad \text{if } |y| \leq \delta.$$

By property b),

$$f_\varepsilon(x) - f(x) = \int f(x-y)K_\varepsilon(y)dy - \int f(x)K_\varepsilon(y)dy =$$

$$= \int [f(x-y) - f(x)]K_\varepsilon(y)\,dy.$$

Hence,

$$|f_\varepsilon(x) - f(x)| \leq \int |f(x-y)-f(x)|\,|K_\varepsilon(y)|dy = \int_{|y|\leq\delta} + \int_{|y|>\delta} = I_1 + I_2.$$

By (1) and property a),

$$I_1 \leq \eta \int_{|y|\leq\delta} |K_\varepsilon(y)|dy \leq \eta \int |K_\varepsilon(y)|dy \leq A\eta.$$

If $|f(x)| \leq M$, then by property c),

$$I_2 \leq 2M \int_{|y|\geq\delta} |K_\varepsilon(y)|dy \to 0 \quad \text{as } \varepsilon \to 0.$$

Therefore,

$$|f_\varepsilon(x) - f(x)| \leq I_1 + I_2 \leq A\eta + \mathcal{O}(1) < 2A\eta$$

if ε is sufficiently small.

The rest of the conclusion follows easily.

Q.E.D.

REMARK: If $|f(x)| \leq M$, then $|f_\varepsilon(x)| \leq AM$.

Moreover, if $K_\varepsilon(x) \geq 0$, then property b) implies a) with $A = 1$.

If $f \in L^p$, $1 \leq p \leq \infty$, then $f_\varepsilon = f*K_\varepsilon$ is also in L^p for all ε, by Th. 2. The next theorem, of which Th. 4 is a sort of limiting case ($p = \infty$), shows that f_ε converges to f in the metric of L^p.

THEOREM 5: If $f \in L^p$, $1 \leq p < \infty$, and $K_\varepsilon(x)$ satisfies conditions a), b) and c), then

$$\|f_\varepsilon - f\|_p \to 0 \quad \text{as } \varepsilon \to 0.$$

Proof: Let $1 < p < \infty$ and $\frac{1}{p} + \frac{1}{p'} = 1$. By Hölder's inequality and properties a) and b),

$$|f_\varepsilon(x) - f(x)| \leq \int |f(x-y) - f(x)|\, |K_\varepsilon(y)|\, dy =$$

$$= \int |f(x-y) - f(x)|\, |K_\varepsilon(y)|^{1/p}\, |K_\varepsilon(y)|^{1/p'} dy \leq$$

$$\leq \left(\int |f(x-y) - f(x)|^p |K_\varepsilon(y)|\, dy\right)^{1/p} \left(\int |K_\varepsilon(y)|\, dy\right)^{1/p'} \leq$$

$$\leq A^{1/p'} \left(\int |f(x-y) - f(x)|^p |K_\varepsilon(y)|\, dy\right)^{1/p}.$$

Hence,

$$\int |f_\varepsilon(x) - f(x)|^p dx \leq A^{p/p'} \int dx \left(\int |f(x-y) - f(x)|^p |K_\varepsilon(y)|\, dy\right)$$

and, interchanging the order of integration, we have

$$\|f_\varepsilon - f\|_p^p \leq A^{p/p'} \int \left(\int |f(x-y) - f(x)|^p dx\right) |K_\varepsilon(y)|\, dy.$$

If we set

$$g(y) = \int |f(x-y) - f(x)|^p dx$$

then g is everywhere defined and bounded by $2\|f\|_p^p$; moreover, by Lemma 1, $g(y) \to 0$ as $y \to 0$. Therefore, by the same argument as in Th. 4, we conclude that

$$\|f_\varepsilon - f\|_p^p \leq A^{p/p'} \int g(y) |K_\varepsilon(y)|\, dy \to 0 \text{ as } \varepsilon \to 0.$$

EXERCISE: Prove Th. 5 in the case $p = 1$.

Suppose that f is a locally integrable function, and let $S(x,\varepsilon) = \{y: |y-x| \leq \varepsilon\}$. By Lebesgue's Theorem on the differenti ability of integrals we have that, as $\varepsilon \to 0$,

$$\frac{1}{|S(x,\varepsilon)|} \int_{S(x,\varepsilon)} f(y)\, dy \to f(x)$$

at almost all points $x \in E^n$. Lemma 3, below, is a strengthening of this result.

DEFINITION: x is a <u>Lebesgue point</u> of f if

$$\frac{1}{|S(x,\varepsilon)|} \int_{S(x,\varepsilon)} |f(y)-f(x)|\, dy \to 0 \text{ as } \varepsilon \to 0.$$

Clearly, any point of continuity of f is also a Lebesgue point of f.

LEMMA 3: Almost every point $x \in E^n$ is a Lebesgue point of f.

Proof: Let a be a rational number. Then, with $S = S(x,\varepsilon)$, as $\varepsilon \to 0$

$$(1) \quad \frac{1}{|S|} \int_S |f(y) - a|\, dy \to |f(x) - a|$$

except possibly for x belonging to a set Z_a of measure zero. Let Z be the union of the sets Z_a, for all rational numbers a; then Z has also measure zero. We show that if $x \notin Z$, then x is a Lebesgue point of f.

Let $\delta > 0$; then, for some rational number a, $|f(x) - a| < \delta/2$.

Now

$$(2) \quad \frac{1}{|S|} \int_S |f(y) - f(x)|\, dy \leq \frac{1}{|S|} \int_S |f(y)-a|\, dy + \frac{1}{|S|} \int_S |a-f(x)|\, dy$$

but the second member of the right-side of (2) is less that $\delta/2$, and the first member tends $|f(x)-a| < \delta/2$, as $\varepsilon \to 0$, by virtue of (1). Hence the left-side of (2) is less then δ, for all sufficiently small ε. Since δ was an arbitrary positive number, it follows that x is a Lebesgue point of f.

Q.E.D.

THEOREM 6: Suppose that K(x) is bounded and that $K(x) = \mathcal{O}(|x|^{-(n+1)})$

for $|x| \geqslant 1$, (hence $K(x)$ is integrable). Suppose also that $\int K(x)\,dx = 1$, and let $K_\varepsilon(x) = \varepsilon^{-n} K(\frac{x}{\varepsilon})$. Then, for any $f \in L^p$, $1 \leqslant p < \infty$,

$$f_\varepsilon(x) = (f*K_\varepsilon)(x) \to f(x) \quad \text{almost everywhere.}$$

Proof: By Lemma 3, it suffices to prove convergence at every Lebesgue point of f. For some positive constant A, we have that

$$|K(y)| \leqslant \frac{A}{1+|y|^{n+1}} \quad \text{for all } y \in E^n;$$

hence,

$$|K_\varepsilon(y)| \leqslant \frac{A\varepsilon}{\varepsilon^{n+1}+|y|^{n+1}} .$$

Let x be a Lebesgue point of f. Since $\int K_\varepsilon(y)\,dy = \int K(y)\,dy = 1$,

$$|f_\varepsilon(x)-f(x)| = \left|\int [f(x-y)-f(x)]K_\varepsilon(y)\,dy\right| \leqslant$$

$$\leqslant \int |f(x-y)-f(x)|\;|K_\varepsilon(y)|\,dy = \int_{|y|\leqslant\varepsilon} + \int_{|y|\geqslant\varepsilon} = I_1 + I_2$$

Now

$$I_1 \leqslant \int_{|y|\leqslant\varepsilon} |f(x-y)-f(x)| \frac{A\varepsilon}{\varepsilon^{n+1}+|y|^{n+1}}\,dy \leqslant \int_{|y|\leqslant\varepsilon} |f(x-y)-f(x)| \frac{A\varepsilon}{\varepsilon^{n+1}}\,dy =$$

$$= \frac{A\varepsilon}{\varepsilon^{n+1}} \int_{|y|\leqslant\varepsilon} |f(x-y)-f(x)|\,dy \to 0 \text{ as } \varepsilon \to 0, \text{ since x is a Lebesgue}$$

point

point of f. On the other hand,

$$I_2 \leqslant \int_{|y|\geqslant\varepsilon} |f(x-y)-f(x)| \frac{A\varepsilon}{\varepsilon^{n+1}+|y|^{n+1}}\,dy \leqslant A\varepsilon \int_{|y|\geqslant\varepsilon} |f(x-y)-f(x)| \frac{dy}{|y|^{n+1}} .$$

Let $F(u) = \int_{|y|\leqslant u} |f(x-y)-f(x)|\,dy.$

Since x is a Lebesgue point of f, $F(u) = \mathcal{O}(u^n)$ as $u \to 0$. Hence, given an arbitrary $\delta > 0$, there exists a(small) $\omega > 0$ such that $F(u) < \delta u^n$

if $u \leq \omega$. Let

$$\varepsilon \int_{|y| \geq \varepsilon} |f(x-y)-f(x)| \frac{dy}{|y|^{n+1}} = \varepsilon \int_{\varepsilon \leq |y| \leq \omega} + \varepsilon \int_{|y| \geq \omega} .$$

The last term clearly tends to zero as $\varepsilon \to 0$. The integral over the "shell" $\varepsilon \leq |y| \leq \omega$ can be treated as a Stieltjes integral; namely

$$\varepsilon \int_{\varepsilon \leq |y| \leq \omega} |f(x-y)-f(x)| \frac{dy}{|y|^{n+1}} = \varepsilon \int_{\varepsilon}^{\omega} \frac{1}{u^{n+1}} dF(u) = \quad \text{(integrating by parts)}$$

$$= \varepsilon \left\{ \left[\frac{F(u)}{u^{n+1}} \right]_{\varepsilon}^{\omega} + (n+1) \int_{\varepsilon}^{\omega} \frac{F(u)}{u^{n+2}} du \right\} \leq$$

$$\leq \varepsilon \frac{F(\omega)}{\omega^{n+1}} + \varepsilon(n+1) \int_{\varepsilon}^{\omega} \frac{\delta u^n}{u^{n+2}} du .$$

As $\varepsilon \to 0$, $\varepsilon F(\omega)\omega^{-n-1}$ tends to zero, because ω is fixed. Finally,

$$\varepsilon(n+1) \int_{\varepsilon}^{\omega} \frac{\delta}{u^2} du < \delta\varepsilon(n+1) \int_{\varepsilon}^{\infty} u^{-2} du = \delta(n+1) ,$$

whence, as $\varepsilon \to 0$, $\limsup I_2 \leq A(n+1)\delta$ for any arbitrarily small δ.

Therefore, at any Lebesgue point of f,

$$|f_{\varepsilon}(x) - f(x)| \leq I_1 + I_2 \to 0 \text{ as } \varepsilon \to 0.$$

Q.E.D.

Since the Poisson Kernel satisfies the hypotheses of Th. 6, we obtain the following useful result from Theorems 5 and 6.

THEOREM 7: The Poisson Kernel, convolved with any $f \in L^P(E^n)$, converges to f almost everywhere and in the L^P norm, $1 \leq P < \infty$.

EXERCISES

1. Prove that any function in L^p, $1 \leq p \leq \infty$, is locally integrable (i.e. integrable over any finite sphere).

2. Suppose that f and g are integrable functions with bounded (hence compact) support. Prove that the convolution $h = f*g$ has bounded support.

3. Consider the convolution $h = f*g$ where f is locally integrable and $g \in C_o^\infty(E^n)$. Prove that $h \in C^\infty(E^n)$, and that for any multi-index α

$$D^\alpha h = f*D^\alpha g.$$

4. Show that Theorem 6 still holds if the condition

$$K(x) = \mathcal{O}(|x|^{-(n+1)})$$

is replaced by the condition

$$K(x) = \mathcal{O}(|x|^{-(n+a)}) \qquad a > 0.$$

CHAPTER II

FOURIER TRANSFORMS

§1. L^1 THEORY

If $x,y \in E^n$ we denote by $x\cdot y$ their inner product, i.e.

$$x\cdot y = x_1y_1 + x_2y_2 + \ldots + x_ny_n.$$

DEFINITION: Let $f \in L(E^n)$. The function $\hat{f}$, given by

$$\hat{f}(x) = \frac{1}{(2\pi)^n}\int_{E^n} f(y)e^{-i(x\cdot y)}dy,$$

is called the Fourier transform of f.

We note that $\hat{f}(x)$ is everywhere defined since the integral is absolutely convergent for every $x \in E^n$. Indeed,

$$|\hat{f}(x)| \leq (2\pi)^{-n}\int |f(y)|dy = (2\pi)^{-n}\|f\|_1,$$

so $\hat{f}(x)$ is bounded. We note also that

$$\hat{f}(0) = (2\pi)^{-n}\int f(y)dy.$$

REMARK: Other common (and essentially equivalent) definitions of Fourier transform are

$$\hat{f}(x) = (2\pi)^{-n/2}\int f(y)e^{-i(x\cdot y)}dy$$

and,

$$\hat{f}(x) = \int f(y)e^{-2\pi i(x\cdot y)}dy.$$

Sometimes the minus sign in the exponential is dropped.

If $n = 1$ and $f \in L(E^1)$, then

$$\hat{f}(x) = \frac{1}{2\pi}\int_{-\infty}^{+\infty} f(y)e^{-ixy}dy.$$

LEMMA 1: If $f(x) = f_1(x_1)f_2(x_2)\ldots f_n(x_n)$, where each $f_j(x_j) \in L(E^1)$,

$$\hat{f}(x) = \hat{f}_1(x_1)\hat{f}_2(x_2)\ldots\hat{f}_n(x_n).$$

Proof: The proof follows immediately from the definition.

It is easy to see that the Fourier transform is a linear transformation, i.e., for any scalars a and b,

$$(af + bg)^{\wedge} = a\hat{f} + b\hat{g}.$$

LEMMA 2: Let $a \in E^n$ and set $f_a(x) = f(x+a)$. Then,

$$\hat{f}_a(x) = e^{i(x.a)}\hat{f}(x).$$

Proof:

$$\hat{f}_a(x) = (2\pi)^{-n} \int f(y+a)e^{-i(x\cdot y)}dy.$$

Letting $y + a = z$, so $y = z-a$ and $dy = dz$, we obtain

$$\hat{f}_a(x) = (2\pi)^{-n} \int f(z)e^{-i[x\cdot(z-a)]}dz =$$

$$= (2\pi)^{-n}e^{i(x\cdot a)} \int f(z)e^{-i(x\cdot z)}dz = e^{i(x\cdot a)}\hat{f}(x).$$

Q.E.D.

LEMMA 3: Let $a \in E^n$ and set $g(x) = e^{i(x\cdot a)}f(x)$. Then,

$$\hat{g}(x) = \hat{f}(x-a).$$

PROOF:

$$\hat{g}(x) = (2\pi)^{-n}\int g(y)e^{-i(x\cdot y)}dy =$$

$$= (2\pi)^{-n}\int g(y)e^{-i[(x-a)\cdot y]}e^{-i(a\cdot y)}dy =$$

$$= (2\pi)^{-n}\int f(y)e^{-i[(x-a)\cdot y]}dy = \hat{f}(x-a).$$

Q.E.D.

LEMMA 4: Let $\lambda \neq 0$ be a real scalar and set $f_\lambda(x) = f(\lambda x)$. Then,

$$\hat{f}_\lambda(x) = \frac{1}{|\lambda|^n}\hat{f}\left(\frac{x}{\lambda}\right).$$

Proof: Using the change of variable $z = \lambda y$, so $dz = |\lambda|^n dy$, we have

$$\hat{f}_\lambda(x) = (2\pi)^{-n}\int f(\lambda y)e^{-i(x\cdot y)}dy =$$

$$= |\lambda|^{-n}(2\pi)^{-n}\int f(z)e^{-i(\frac{x}{\lambda}.z)}dz = |\lambda|^{-n}\hat{f}(\tfrac{x}{\lambda})$$

Q.E.D.

We recall that a function f is said to be <u>even</u> if $f(-x) = f(x)$, and is said to be <u>odd</u> if $f(-x) = -f(x)$. Moreover, any f can be decomposed into its even and odd parts; namely

$$f(x) = \frac{f(x) + f(-x)}{2} + \frac{f(x) - f(-x)}{2} .$$

LEMMA 5: The Fourier transform of an even function is even. The Fourier transform of an odd function is odd.

Proof: The proof follows directly from Lemma 4, taking $\lambda = -1$.

Q.E.D.

EXAMPLE 1: Let $\chi_h(x)$ be the characteristic function of the interval $(-h,h)$. Then

$$\hat{\chi}_h(x) = \frac{1}{\pi}\,\frac{\sin hx}{x} .$$

In fact,

$$\hat{\chi}_h(x) = \frac{1}{2\pi}\int_{-h}^{+h} e^{-ixy}dy = \frac{1}{2\pi}\,\frac{e^{-ixy}}{-ix}\Bigg|_{-h}^{+h} = \frac{1}{2\pi}\,\frac{e^{ixh}-e^{-ixh}}{ix} =$$

$$= \frac{1}{\pi}\,\frac{\sin hx}{x} .$$

Thus $\hat{\chi}_h$ is a bounded continuous function which is not integrable on E^1, though it belongs to $L^p(E^1)$ for all $p > 1$. Moreover $\hat{\chi}_h(x) \to 0$ as $|x| \to \infty$.

EXAMPLE 2: Let $\chi_{a,b}(x)$ be the characteristic function of the interval (a,b). Letting $c = \frac{a+b}{2}$ and $h = \frac{b-a}{2}$, we have that

$$\chi_{a,b}(x) = \chi_h(x-c).$$

Hence, by EXAMPLE 1 and LEMMA 2,

$$\hat{\chi}_{a,b}(x) = \frac{1}{\pi} \frac{\sin hx}{x} e^{-icx}.$$

We note again that $\hat{\chi}_{a,b}(x) \to 0$ as $|x| \to \infty$. The next theorem shows that this is a general property of Fourier transforms of integrable functions.

THEOREM 1: (Riemann-Lebesgue) If $f \in L(E^n)$ then $\hat{f}$ is bounded and uniformly continuous; moreover,

$$(*) \qquad \hat{f}(x) \to 0 \text{ as } |x| \to \infty.$$

Proof: As we saw earlier, $|\hat{f}(x)| \leq (2\pi)^{-n} \|f\|_1$, so $\hat{f}$ is a bounded function.

Let $\delta > 0$ and $h = (h_1, \ldots, h_n) \in E^n$; then, for any $x \in E^n$,

$$|\hat{f}(x+h) - \hat{f}(x)| = |(2\pi)^{-n} \int f(y)[e^{-i(x+h)\cdot y} - e^{-i(x\cdot y)}]dy| =$$

$$= |(2\pi)^{-n} \int f(y) e^{-i(x\cdot y)}[e^{-i(h\cdot y)} - 1]dy| \leq$$

$$\leq \int |f(y)| |e^{-i(h\cdot y)} - 1| dy = \int_{|y|\leq R} + \int_{|y|>R} = I_1 + I_2.$$

Now,

$$I_2 \leq 2 \int_{|y|>R} |f(y)| dy < \delta$$ if R is large enough (but fixed), since f is integrable.

On the other hand, if $|y| \leq R$ then $|e^{-i(h\cdot y)} - 1| \to 0$ as $|h| \to 0$, so

$$I_1 \leq \int_{|y|\leq R} |f(y)| \mathcal{O}(1) dy < \delta$$ if $|h|$ is small enough.

Thus $\hat{f}$ is uniformly continuous.

Finally we prove (*) in a series of steps.

(i) If $f = g + h$, where g satisfies (*) and $\|h\|_1$ is arbitrarily small, then f satisfies (*). In fact $\hat{f} = \hat{g} + \hat{h}$ and $\hat{g}(x) \to 0$ as $|x| \to \infty$, $|\hat{h}(x)| \leq (2\pi)^{-n} \|h\|_1$ is small.

(ii) Characteristic functions of one-dimensional intervals satisfy (*), as we noted in EXAMPLE 2 above.

(iii) Characteristic functions of n-dimensional intervals $a_i \leq x_i \leq b_i$, $i = 1,\dots,n$, satisfy (*).

This follows directly from (ii) and LEMMA 1.

(iv) Simple functions, i.e. finite linear combinations of functions in (iii) satisfy (*).

(v) Simple functions are dense in $L(E^n)$; so, by (i), we are done,

Q.E.D.

In view of the presence of the factor $(2\pi)^{-n}$ in the definition of Fourier transform, it is convenient to normalize accordingly the definition of convolution. So if $f, g \in L(E^n)$ we set

$$h(x) = (f*g)(x) = \frac{1}{(2\pi)^n}\int f(y)g(x-y)\,dy.$$

We know that $h \in L$ also, hence $\hat{h}$ exists. The next theorem illustrates further the relationship between convolution and multiplication of functions.

THEOREM 2: Let $f, g \in L$. If $h = f*g$ then $\hat{h} = \hat{f}\hat{g}$.

Proof:

$$\hat{h}(x) = (2\pi)^{-n}\int h(y)e^{-i(x\cdot y)}\,dy = (2\pi)^{-n}\int [(2\pi)^{-n}\int f(z)g(y-z)\,dz]e^{-i(x\cdot y)}\,dy$$

exists for every x. Using the identity $e^{-i(x\cdot y)} = e^{-i[x\cdot(y-z)]}e^{-i(x\cdot z)}$ and interchanging the order of integration, we obtain

$$\hat{h}(x) = (2\pi)^{-n}\int f(z)e^{-i(x\cdot z)}[(2\pi)^{-n}\int g(y-z)e^{-i[x\cdot(y-z)]}dy]dz =$$

$$= [(2\pi)^{-n}\int f(z)e^{-i(x\cdot z)}dz][(2\pi)^{-n}\int g(w)e^{-i(x\cdot w)}dw] = \hat{f}(x)\hat{g}(x).$$

Q.E.D.

EXAMPLE 3: Let $x \in E^1$ and $0 < k \leq h$. Consider the "trapezoidal function" $\chi_{h,k}(x)$ defined as follows: $\chi_{h,k}(x)$ is even, and

$$\chi_{h,k}(x) = \begin{cases} 1 & \text{if } 0 \leq x \leq h - k \\ 0 & \text{if } x \geq h + k \\ \dfrac{h+k-x}{2k} & \text{if } x \in (h-k, h+k) \end{cases}$$

This function is essentially the convolution of the characteristic functions χ_h and χ_k. In fact, it is easy to verify that

$$(\chi_h * \chi_k)(x) = \frac{1}{2\pi} \int \chi_h(t)\chi_k(x-t)\,dt = \frac{1}{2\pi} \int_{-h}^{+h} \chi_k(x-t)\,dt = \frac{2k}{2\pi} \chi_{h,k}(x).$$

So,

$$\chi_{h,k} = \frac{\pi}{k} \chi_h * \chi_k \text{ and } \hat{\chi}_{h,k} = \frac{\pi}{k} \hat{\chi}_h \hat{\chi}_k; \text{ therefore,}$$

from EXAMPLE 1, it follows that the Fourier transform of $\chi_{h,k}$ is

(1) $$\hat{\chi}_{h,k}(x) = \frac{1}{\pi k} \frac{\sin hx}{x} \frac{\sin kx}{x}, \quad 0 < k \leq h.$$

In particular, if $k = h$, we have the "triangular function"

$$\chi_{h,h}(x) = \left(1 - \frac{|x|}{2h}\right)^+$$

where $f^+(x) = \max[f(x), 0]$ is the positive part of $f(x)$. So, from formula (1), we have

(2) $$\hat{\chi}_{h,h}(x) = \frac{1}{\pi h} \frac{\sin^2 hx}{x^2}, \quad h > 0.$$

We note that $\hat{\chi}_{h,h}(x)$ is non-negative and integrable on E^1.

EXAMPLE 4: Let $H(x) = e^{-|x|}$, $-\infty < x < \infty$. Then

$$\hat{H}(x) = \frac{1}{\pi} \cdot \frac{1}{1+x^2}.$$

In fact

$$\hat{H}(x) = \frac{1}{2\pi}\int_{0}^{\infty} e^{-y}e^{-ixy}dy + \frac{1}{2\pi}\int_{-\infty}^{0} e^{y}e^{-ixy}dy =$$

$$= \frac{1}{2\pi}\int_{0}^{\infty} e^{-y(1+ix)}dy + \frac{1}{2\pi}\int_{-\infty}^{0} e^{y(1-ix)}dy =$$

$$= \frac{1}{2\pi}\left(\frac{1}{1+ix} + \frac{1}{1-ix}\right) = \frac{1}{\pi}\cdot\frac{1}{1+x^2}$$

Moreover, if $\varepsilon > 0$ and $H_\varepsilon(x) = H(\varepsilon x) = e^{-\varepsilon|x|}$, then by Lemma 4,

$$\hat{H}_\varepsilon(x) = \frac{1}{\pi}\cdot\frac{\varepsilon}{\varepsilon^2+x^2}$$

which is the Poisson Kernel in E^1.

EXAMPLE 5: Let $H(x) = e^{-x^2}$, $-\infty < x < \infty$. Then

$$\hat{H}(x) = (2\sqrt{\pi})^{-1}e^{-x^2/4}.$$

In fact,

$$\hat{H}(y) = (2\pi)^{-1}\int_{-\infty}^{\infty} e^{-x^2}e^{-ixy}dx = \text{(completing squares)}$$

$$= (2\pi)^{-1}\int_{-\infty}^{\infty} e^{-[x^2+ixy+\left(\frac{iy}{2}\right)^2 - \left(\frac{iy}{2}\right)^2]}dx =$$

$$= (2\pi)^{-1}e^{-y^2/4}\int_{\infty}^{\infty} e^{-(x+\frac{iy}{2})^2}dx = (2\pi)^{-1}e^{-y^2/4}\int_{L} e^{-z^2}dz$$

where L is the line $\mathrm{Im}(z) = y/2$. By a simple application of Cauchy's Theorem we have that

$$\int_{L} e^{-z^2}dz = \int_{-\infty}^{\infty} e^{-x^2}dx = \sqrt{\pi}.$$

Hence,

$$\hat{H}(y) = (2\sqrt{\pi})^{-1} e^{-y^2/4}$$

Again, if $\varepsilon > 0$ and $H_\varepsilon(x) = H(\varepsilon x) = e^{-\varepsilon^2 x^2}$, then

$$\hat{H}_\varepsilon(x) = (2\varepsilon\sqrt{\pi})^{-1} e^{-x^2/4\varepsilon^2}$$

which is the Weierstrass Kernel in E^1. In particular, taking $\varepsilon = 2^{-1/2}$, we see that the function $e^{-x^2/2}$ is equal to its Fourier transform, up to the constant factor $(2\pi)^{-1/2}$.

THEOREM 3: Let $f, g \in L$; then

$$\int \hat{f}(x) g(x) dx = \int f(x) \hat{g}(x) dx.$$

Proof: By Th. 1, $\hat{f}g$ and $f\hat{g}$ are integrable, so the above integrals exist. Now, interchanging the order of integration,

$$\begin{aligned}\int \hat{f}(x) g(x) dx &= \int g(x) \{(2\pi)^{-n} \int f(y) e^{-i(x\cdot y)} dy\} dx = \\ &= \int f(y) \{(2\pi)^{-n} \int g(x) e^{-i(x\cdot y)} dx\} dy = \\ &= \int f(y) \hat{g}(y) dy.\end{aligned}$$

Q.E.D.

We now pass to consider the effect of an invertible linear transformation of E^n on the Fourier transform of a function. Let $T: E^n \to E^n$ be an invertible linear transformation, so $\det T \neq 0$, and denote by T^{-1} and T' the inverse and the transpose of T. We set $f_T(x) = f(Tx)$ and calculate the Fourier transform of this function.

If $Ty = z$ then $y = T^{-1}z$ and $dz = |\det T| dy$, so

$$\hat{f}_T(x) = (2\pi)^{-n} \int f(Ty) e^{-i(x\cdot y)} dy =$$

$$= (2\pi)^{-n}|\det T|^{-1} \int f(Ty)e^{-i(x\cdot y)}|\det T|\,dy =$$

$$= (2\pi)^{-n}|\det T|^{-1} \int f(z)e^{-i(x\cdot T^{-1}z)}dz =$$

$$= (2\pi)^{-n}|\det T|^{-1} \int f(z)e^{-i[(T^{-1})'x\cdot z]}dz$$

hence,

(3) $$\hat{f}_T(x) = |\det T|^{-1}\hat{f}((T^{-1})'x).$$

DEFINITION: f is a <u>radial</u> function if $f(x) = f(|x|)$.

Equivalently, a radial function is invariant under all rotations about the origin. For example $|x|^\alpha, e^{|x|^\alpha}$, etc. are all radial functions.

THEOREM 4: The Fourier transform of a radial function is radial.

Proof: For any orthogonal transformation T, we have that $|\det T| = 1$ and $T = (T^{-1})'$, thus formula (3) becomes

(3') $\hat{f}_T(x) = \hat{f}(Tx)$, T orthogonal.

Hence the Fourier transform commutes with orthogonal transformations. Now, f is radial if and only if $f(Tx) = f(x)$, for all orthogonal transformations T, therefore it follows from (3') that

$$\hat{f}(x) = \hat{f}_T(x) = \hat{f}(Tx)$$

so $\hat{f}$ is also radial.

Q.E.D.

§2. THE INVERSION FORMULA

If f(x) is an integrable function, periodic of period 2π, and

$$c_n = \hat{f}(n) = \frac{1}{2\pi}\int_{-\pi}^{\pi} f(y)e^{-iny}dy$$

are its Fourier coefficients, then, in some cases, we can represent f(x) by means of its Fourier series:

(i) $$f(x) = \sum_{-\infty}^{+\infty} \hat{f}(n)e^{inx} .$$

In general the Fourier series does not converge to the values of the function, but it is "summable", in some sense, to these values. Similarly, if $f \in L(E^n)$ and

$$\hat{f}(y) = (2\pi)^{-n}\int f(x)e^{-i(x\cdot y)}dx$$

is it Fourier transform, we seek a representation of f(x) by means of its Fourier transform:

(ii) $f(x) = \int \hat{f}(y)e^{i(x\cdot y)}dy$ (INVERSION FORMULA).

However $\hat{f}$ need not be integrable (cf. EXAMPLE 1 of §1) so the integral in (ii) is not convergent in general. For this reason we introduce now the concept of summability of integrals.

Let $a(u)$, $0 \leq u < \infty$, be a locally integrable function so that the "partial integrals"

$$A(\omega) = \int_0^{\omega} a(u)du$$

exist. But

$$\int_0^{\infty} a(u)du = \lim_{\omega\to\infty} A(\omega)$$

need not exist; in other words, the integral need not converge.

Consider a function $H(u)$, $0 \leq u < \infty$, satisfying the following conditions:

1) $H(u)$ is of bounded variation (b.v.) on $[0,\infty)$;

2) $H(u) \to 0$ as $u \to \infty$;

3) $H(u)$ is continuous at $u = 0$ and $H(0) = 1$.

DEFINITION: The integral $\int_0^\infty a(u)du$ is said to be summable - H to a value I if the integrals

$$I_\varepsilon = \int_0^\infty a(u)H(\varepsilon u)du \text{ converge for all } \varepsilon > 0$$

and $\lim_{\varepsilon\to 0} I_\varepsilon = I$.

EXAMPLES:

1. Let
$$H(u) = \begin{cases} 1 & \text{if } 0 \leq u \leq 1 \\ 0 & \text{if } u > 1. \end{cases}$$

Clearly $H(u)$ satisfies the previous three conditions; moreover

$$I_\varepsilon = \int_0^\infty a(u)H(\varepsilon u)du = \int_0^{1/\varepsilon} a(u)du$$

thus, in this case, summability corresponds to ordinary convergence.

2. Let $H(u) = (1-u)^+ = \max[1-u,0]$, where $u \geq 0$. Again $H(u)$ has the desired properties and, letting $\varepsilon = 1/\omega$ so $\varepsilon \to 0$ as $\omega \to \infty$, we have that

$$I_\varepsilon = \int_0^\infty a(u)H(\varepsilon u)du = \int_0^{1/\varepsilon} a(u)(1-\varepsilon u)du = \int_0^\omega a(u)(1-\frac{u}{\omega})du.$$

This is called the method of summability by first arithmetic means; in fact if

$$A(v) = \int_0^v a(u)\,du \quad \text{and} \quad B(\omega) = \frac{1}{\omega}\int_0^\omega A(v)\,dv$$

then

$$B(\omega) = \frac{1}{\omega}\int_0^\omega dv\left(\int_0^v a(u)\,du\right) = \text{interchanging the order of integration}$$

$$= \frac{1}{\omega}\int_0^\omega a(u)\left(\int_u^\omega 1\,dv\right)du =$$

$$= \frac{1}{\omega}\int_0^\omega a(u)(\omega-u)\,du = \int_0^\omega a(u)\left(1-\frac{u}{\omega}\right)du$$

as before.

3. The function $H(u) = \begin{cases} (1-u)^r & \text{if } 0 \leqslant u \leqslant 1 \\ 0 & \text{if } u > 1 \end{cases}$

and $H(u) = \begin{cases} (1-u^2)^r & \text{if } 0 \leqslant u \leqslant 1 \\ 0 & \text{if } u > 1 \end{cases}$

where $r > 0$, both define the method of summability by <u>r^{th} arithmetic means</u>.

4. The function $H(u) = e^{-u}$, $u \geqslant 0$, defines the <u>Abel-Poisson method</u> of summability. Letting $e^{-\varepsilon} = r$, then $r < 1$ and $r \to 1$ as $\varepsilon \to 0$; moreover

$$I_\varepsilon = \int_0^\infty a(u)H(\varepsilon u)\,du = \int_0^\infty a(u)e^{-\varepsilon u}\,du = \int_0^\infty a(u)r^u\,du.$$

5. The function $H(u) = e^{-u^2}$, $u \geqslant 0$, defines the <u>Gauss-Weierstrass method</u> of summability. Here,

$$I_\varepsilon = \int_0^\infty a(u)e^{-\varepsilon^2u^2}\,du.$$

The following theorem justifies the definition of summability.

THEOREM 5: Let a(u) be locally integrable on $[0,\infty)$. If the integral

$$\int_0^\infty a(u)\,du$$

converges to a value I, then it is summable - H to the value I.

Proof: We must show that $I_\varepsilon = \int_0^\infty a(u)H(\varepsilon u)\,du$ exists, and that $I_\varepsilon \to I$, as $\varepsilon \to 0$.

By our assumption, the function $A(u) = \int_0^u a(v)\,dv$ exists and $A(u) \to I$ as $u \to \infty$. Moreover,

$$A'(u) = a(u) \quad \text{almost everywhere.}$$

We denote by V the total variation of H(u) on $[0,\infty)$. Integrating by parts, we have that

$$\int_0^\omega a(u)H(\varepsilon u)\,du = \int_0^\omega H(\varepsilon u)A'(u)\,du = \int_0^\omega H(\varepsilon u)\,dA(u) =$$

$$= H(\varepsilon\omega)A(\omega) - \int_0^\omega A(u)\,dH(\varepsilon u)$$

where the last integral exists since, if $|A(u)| \leq M$,

$$\left|\int_0^\omega A(u)\,dH(\varepsilon u)\right| \leq MV$$

because H(u) and $H(\varepsilon u)$ have the same total variation. Letting $\omega \to \infty$, we obtain

$$I_\varepsilon = \int_0^\infty a(u)H(\varepsilon u)\,du = -\int_0^\infty A(u)\,dH(\varepsilon u)$$

since $H(\varepsilon\omega) \to 0$ and $A(\omega) \to I$.

If we set $A(u) = I + \eta(u)$, so $\eta(u) \to 0$ as $u \to \infty$, then

$$I_\varepsilon = -I \int_0^\infty dH(\varepsilon u) - \int_0^\infty \eta(u)\, dH(\varepsilon u) =$$

$$= -I[H(\infty) - H(0)] - \int_0^\infty \eta(u)\, dH(\varepsilon u) =$$

$$= I - \int_0^\infty \eta(u)\, dH(\varepsilon u)$$

and it remains to show that this last integral tends to zero as $\varepsilon \to 0$.

By our assumption $|\eta(u)| \leqslant N$ and, for any given $\delta > 0$, $|\eta(u)| \leqslant \delta$ if $u \geqslant u_o$. Now,

$$\int_0^\infty \eta(u)\, dH(\varepsilon u) = \int_0^{u_o} + \int_{u_o}^\infty = P + Q$$

and

$|Q| \leqslant \delta V$ is arbitrarily small, for all ε. On the other hand,

$$|P| \leqslant N \text{ (variation of } H(\varepsilon u) \text{ on } [0,u_o]) =$$

$$= N \text{ (variation of } H(u) \text{ on } [0,\varepsilon u_o]) \to 0 \text{ as } \varepsilon \to 0$$

since $H(u)$ is of b.v. and continuous at $u = 0$, hence its variation is also continuous there.

Q.E.D.

We now extend the concept of summability to multiple integrals. Let $a(u)$ be a locally integrable function of $u \in E^n$, and

$$\int a(u)\, du = \int \cdots \int a(u_1, \ldots, u_n)\, du_1 \ldots du_n .$$

Here it is most natural to consider partial integrals taken over spheres with center at the origin; namely,

$$S(R) = \int_{|u| \leqslant R} a(u)\, du .$$

Then, we say that $\int a(u)\,du$ converges spherically to a value I if $S(R) \to I$ as $R \to \infty$.

The corresponding notion of spherical summability is defined as follows: we consider integrals

$$I_\varepsilon = \int a(u) H(\varepsilon|u|)\,du$$

where H(t) satisfies conditions 1), 2) and 3) listed earlier; if I_ε exists for every $\varepsilon > 0$ and $I_\varepsilon \to I$ as $\varepsilon \to 0$, then we say that $\int a(u)\,du$ is summable - H to the value I.

We note that $H(|u|)$ is a radial function on E^n. If $H(|u|) = 1$, for $|u| \leq 1$, and is equal to zero elsewhere, then we have spherical convergence. If $H(|u|) = e^{-|u|}$ we have the Abel-Poisson summability; the function $H(|u|) = e^{-|u|^2}$ defines the Gauss-Weierstrass summability and so on.

THEOREM 6: Let a(u) be a locally integrable function on E^n. If the integral $\int a(u)\,du$ converges spherically to a value I, then it is also summable - H to the same value I.

Proof: By hypothesis, $S(R) = \int_{|u| \leq R} a(u)\,du \to I$ as $R \to \infty$.

Integrating by parts,

$$\int_{|u| \leq \omega} a(u) H(\varepsilon|u|)\,du = \int_0^\omega H(\varepsilon R)\,dS(R) =$$

$$= H(\varepsilon\omega) S(\omega) - \int_0^\omega S(R)\,dH(\varepsilon R)$$

and, letting $\omega \to \infty$, we have that

$$I_\varepsilon = \int a(u) H(\varepsilon|u|)\,du = - \int_0^\infty S(R)\,dH(\varepsilon R).$$

The rest of the proof is identical to that of Th. 5.

Q.E.D.

We now return to the Inversion Formula,

$$(*) \qquad f(x) = \int \hat{f}(y) e^{i(x\cdot y)} dy, \qquad f \in L(E^n).$$

We will show that, under some additional assumptions on the function $H(|x|)$, the integral in (*) is summable - H to f(x) for almost every x.

We recall that, with $r = |x|$, $H(r)$ satisfies conditions

1) $H(r)$ is of b.v. on $[0,\infty)$,

2) $H(r) \to 0$ as $r \to \infty$,

3) $H(r)$ is continuous at $r = 0$ and $H(0) = 1$.

Assume, in addition, that

4) $H(|x|) \in L(E^n)$.

In other words, using polar coordinates,

$$\int H(|x|)\, dx = C \int_0^\infty H(r) r^{n-1}\, dr < \infty.$$

If we denote by $K(x)$ the Fourier transform of $H(|x|)$, then $K(x)$ is a bounded radial function. By Lemma 2 and Th. 3, we have that

$$\int \hat{f}(y) e^{i(x\cdot y)} H(\varepsilon|y|)\, dy = \int \widehat{f(y+x)} H(\varepsilon|y|)\, dy =$$

$$= \int f(y+x) \widehat{H(\varepsilon|y|)} dy = \int f(x-y) \widehat{H(\varepsilon|y|)} dy.$$

But, by Lemma 4, $\widehat{H(\varepsilon|y|)} = \varepsilon^{-n}\hat{H}\left(\frac{|y|}{\varepsilon}\right) = \varepsilon^{-n}K\left(\frac{y}{\varepsilon}\right) = K_\varepsilon(y)$; hence,

$$\int \hat{f}(y) e^{i(x\cdot y)} H(\varepsilon|y|) dy = \int f(x-y) K_\varepsilon(y)\, dy = (f*K_\varepsilon)(x).$$

Recalling that $K(x)$ is bounded, we assume finally that

5) $K(x) = \hat{H}(|x|) = \mathcal{O}(|x|^{-n-1})$ for $|x| \geq 1$, (hence $K(x)$ is integrable),

6) $\int K(x)\, dx = 1$.

In view of Theorems 5 and 6 of Chapter I, we obtain the following conclusion:

THEOREM 7: If $f \in L(E^n)$ and $H(|x|)$ satisfies conditions 1) through 6) above, then, as $\varepsilon \to 0$,

$$\int \hat{f}(y) e^{i(x\cdot y)} H(\varepsilon|y|)dy \to f(x)$$

almost everywhere and in L^1 norm.

COROLLARY: (Uniqueness of the Fourier transform) If $\hat{f}(x) \equiv 0$, then $f(x) = 0$ a.e.

More precisely, we know that under our assumptions

$$\int \hat{f}(y) e^{i(x\cdot y)} H(\varepsilon|y|)\, dy = (f*K_\varepsilon)(x) \to f(x) \quad \text{as } \varepsilon \to 0,$$

at every Lebesgue point of f, and in particular at every point of continuity of f. Hence Theorems 6 and 7 imply

THEOREM 8: If the integral $\int \hat{f}(y) e^{i(x\cdot y)}\, dy$ converges (spherically), then it converges a.e. to the value $f(x)$. If $f(x)$ is continuous at some point x_o and $\int \hat{f}(y) e^{i(x\cdot y)}\, dy$ converges at x_o, then it must converge to $f(x_o)$.

In particular, if $f \in L$ and $\hat{f} \in L$ also, then the Inversion Formula

$$(*) \qquad f(x) = \int \hat{f}(y) e^{i(x\cdot y)}\, dy$$

holds almost everywhere; if, in addition, f is continuous then (*) holds everywhere.

In general, if $f \in L(E^n)$, the Inversion Formula is valid a.e. in the sense of Gauss-Weierstrass summability, for example. In fact, the function $H(|x|) = e^{-|x|^2} = e^{-(x_1^2+\ldots+x_n^2)}$ clearly satisfies conditions 1) through 4). Moreover by Lemma 1 and Example 5 of §1,

$$K(x) = \widehat{H(|x|)} = \widehat{e^{-x_1^2}} \dots \widehat{e^{-x_n^2}} = (2\sqrt{\pi})^{-n} e^{-\frac{x_1^2}{4}} \dots e^{-\frac{x_n^2}{4}} =$$

$$= (2\sqrt{\pi})^{-n} e^{-\frac{|x|^2}{4}}$$

so condition 5) is also satisfied. Finally,

$$\int_{E^n} K(x)dx = (2\sqrt{\pi})^{-n} \int_{E^n} e^{-\frac{x_1^2}{4}} \dots e^{-\frac{x_n^2}{4}} dx_1 \dots dx_n =$$

$$= \left[\frac{1}{2\sqrt{\pi}} \int_{-\infty}^{+\infty} e^{-u^2/4} du\right]^n = \left[\frac{1}{\sqrt{\pi}} \int_{-\infty}^{+\infty} e^{-v^2} dv\right]^n = 1^n = 1$$

thus condition 6) holds.

We recall from Example 3 of §1 that the triangular function

$$\chi_{h,h}(x) = \left(1 - \frac{|x|}{2h}\right)^+ , \quad h > 0,$$

is continuous and integrable on E^1, and that the same is true of its Fourier transform

$$\hat{\chi}_{h,h}(x) = \frac{1}{\pi h} \frac{\sin^2 hx}{x^2} .$$

Hence, by the Inversion Formula, we have that for every x

$$\frac{1}{\pi h} \int_{-\infty}^{\infty} \frac{\sin^2 hy}{y^2} e^{ixy} dy = \chi_{h,h}(x).$$

In particular, at x = 0

$$\frac{1}{\pi h} \int_{-\infty}^{\infty} \frac{\sin^2 hy}{y^2} dy = 1$$

and, letting h = 1, we obtain

$$\frac{1}{\pi}\int_{-\infty}^{\infty}\left(\frac{\sin y}{y}\right)^2 dy = 1.$$

It is now easily verified that the function $H(|x|) = \left(1 - \frac{|x|}{2}\right)^+$ satisfies conditions 1) through 6) on E^1.

Since the integral

$$\int_{-\infty}^{\infty} a(u)\, du$$

is summable by the method of first arithmetic means to a value I if

$$I = \lim_{\omega\to\infty}\left[\frac{1}{\omega}\int_0^{\omega} du \int_{-u}^{u} a(v)\,dv\right] = \lim_{\omega\to\infty}\left[\frac{1}{\omega}\int_{-\omega}^{\omega} a(v)\,dv \int_{|v|}^{\omega} du\right] =$$

$$= \lim_{\omega\to\infty}\int_{-\omega}^{\omega} a(v)\left(1 - \frac{|v|}{\omega}\right)dv = \lim_{\omega\to\infty}\int_{-\infty}^{\infty} a(v)\ \left(1 - \frac{|v|}{\omega}\right)^+ dv$$

and since

$$\lim_{\omega\to\infty}\int_{-\infty}^{\infty}\hat{f}(y)e^{ixy}\left(1 - \frac{|y|}{\omega}\right)^+ dy = \lim_{\varepsilon\to\infty}\int_{-\infty}^{\infty}\hat{f}(y)e^{ixy}H(\varepsilon|y|)dy$$

where $\varepsilon = 2/\omega$ and $H(|y|) = \left(1 - \frac{|y|}{2}\right)^+$, we obtain from Th. 7 the following result:

if $f \in L(E^1)$, the integral $\int_{-\infty}^{\infty}\hat{f}(y)e^{ixy}\,dy$

is summable to f(x) a.e. and in L^1 norm, by the method of first arithmetic means.

EXERCISE: If $f \in L(E^1)$, show that the integral $\int_{-\infty}^{\infty}\hat{f}(y)e^{ixy}\,dy$ is summable to f(x) a.e. and in L^1 norm, by the Abel-Poisson method. (Hint: Use Example 4 of §1, and Th. 7 of Chapter I).

REMARK: In view of the inversion Formula, the transformation

$$\check{g}(x) = \int_{E^n} g(y) e^{i(x \cdot y)} \, dy$$

is usually called the <u>Inverse Fourier transform</u>.

§3. L^2 THEORY

We recall that $L^2 = L^2(E^n)$ is a Hilbert space, with inner product

$$(f,g) = \int f(x)\,\overline{g(x)}\,dx$$

which is finite by Schwarz's Inequality (i.e., Hölder's Inequality with $p = 2$), and with norm

$$\|f\|_2 = (f,f)^{1/2} = \left(\int |f(x)|^2\right)^{1/2}.$$

We denote by S the class of all simple functions (i.e. finite linear combinations of characteristic functions of n-dimensional intervals) and recall that S is a dense subspace of L^2.

If $f \in L^2$ then, by Schwarz's Inequality, f is locally integrable, and hence the following "truncated" integrals exist:

$$\hat{f}_R(x) = (2\pi)^{-n} \int_{|y| \leq R} f(y)\,e^{-i(x\cdot y)}\,dy.$$

THEOREM 9: (Parseval-Plancherel) Let $f \in L^2(E^n)$; then the Fourier transform

$$\hat{f}(x) = (2\pi)^{-n} \int f(y)\,e^{-i(x\cdot y)}\,dy$$

exists as a limit in L^2 norm of the $\hat{f}_R(x)$, $R \to \infty$. Also,

$$\|\hat{f}\|_2 = (2\pi)^{-n/2}\,\|f\|_2 \qquad \text{(Parseval Formula)}.$$

Moreover the Inversion Formula

$$f(x) = \int \hat{f}(y)\,e^{i(x\cdot y)}\,dy$$

holds in the sense that

$$f(x) = \underset{R \to \infty}{\text{limit in } L^2} \int_{|y| \leq R} \hat{f}(y)\,e^{i(x\cdot y)}\,dy.$$

Proof: We shall first prove the theorem for simple functions and then pass to the limit. We do so in the following series of steps.

Step 1: Characteristic functions on E^1.

Recall that toward the end of §2 we proved the formula

$$\int_{-\infty}^{\infty} \left(\frac{\sin x}{x}\right)^2 dx = \pi .$$

If $\chi_h(x)$ is the characteristic function of the interval $(-h,h)$, then its Fourier transform

$$\hat{\chi}_h(x) = \frac{1}{\pi}\frac{\sin hx}{x}$$

exists pointwise and also in the L^2 sense, since, with $f(x) = \chi_h(x)$, the truncated integrals

$$\hat{f}_R(x_o) = \frac{1}{2\pi}\int_{-R}^{R} f(y)e^{-ix_oy}\,dy$$

are all the same for $R \geqslant h$. Moreover,

$$\int_{-\infty}^{\infty} \hat{\chi}_h^2(x)\,dx = \frac{1}{\pi^2}\int_{-\infty}^{\infty}\frac{\sin^2 hx}{x^2}\,dx = \text{(letting } y = hx)$$

$$= \frac{h}{\pi^2}\int_{-\infty}^{\infty}\frac{\sin^2 y}{y^2}\,dy = \frac{h}{\pi} =$$

$$= \frac{2h}{2\pi} = \frac{1}{2\pi}\int_{-\infty}^{\infty}\chi_h^2(x)\,dx$$

whence, taking square roots, we see that Parseval's Formula holds.

If $\chi(x)$ is the characteristic function of any interval $(a,b) = (c-h, c+h)$, then we know that

$$\hat{\chi}(x) = \hat{\chi}_h(x)e^{-icx}$$

so $|\hat{\chi}(x)|^2 = |\hat{\chi}_h(x)|^2 = \hat{\chi}_h^2(x)$ and the conclusion follows from the previous argument.

Step 2: Characteristic functions on E^n.

If $\chi(x)$ is the characteristic function of an n-dimensional interval $a_j \leqslant x_j \leqslant b_j$, $j = 1,\dots,n$, then

$$\chi(x) = \chi_1(x_1) \dots \chi_n(x_n)$$

where $\chi_j(x_j)$ is the characteristic function of the 1-dimensional interval (a_j, b_j); also $\hat{\chi}(x)$ is defined in the usual way and

$$\hat{\chi}(x) = \hat{\chi}_1(x_1) \dots \hat{\chi}_n(x_n).$$

The formula

$$\int_{E^n} |\hat{\chi}(x)|^2 \, dx = (2\pi)^{-n} \int_{E^n} |\chi(x)|^2 \, dx$$

follows directly from the previous case since the integrals are just products of one-dimensional integrals of functions in Step 1.

Step 3: Simple functions on E^1

Suppose that $f(x)$ takes values $c_1,\dots,c_m$ on m non-overlapping intervals $I_1,\dots,I_m$, with characteristic functions $\chi_1,\dots,\chi_m$, and $f(x) = 0$ elsewhere. So

$$f(x) = \sum_{k=1}^{m} c_k \chi_k(x)$$

and

$$\hat{f}(x) = \sum_{k=1}^{m} c_k \hat{\chi}_k(x).$$

Then,

$$\int_{-\infty}^{\infty} |\hat{f}(x)|^2 dx = \int_{-\infty}^{\infty} \hat{f}(x)\overline{\hat{f}(x)}\,dx = \int_{-\infty}^{\infty} \left\{ \sum_{k=1}^{m} c_k \hat{\chi}_k(x) \right\} \left\{ \sum_{l=1}^{m} \bar{c}_l \overline{\hat{\chi}_l(x)} \right\} dx =$$

$$= \sum_{k=1}^{m} |c_k|^2 \int_{-\infty}^{\infty} |\hat{\chi}_k(x)|^2 dx + \sum_{k \neq l} c_k \bar{c}_l \int_{-\infty}^{\infty} \hat{\chi}_k(x)\overline{\hat{\chi}_l(x)}\, dx =$$

$= A + B$

CLAIM: $B = 0$, since each term vanishes.

In fact, if $\chi_1(x)$ is the characteristic function $I_1 = (x_1-h_1,\ x_1+h_1)$ and $\chi_2(x)$ is the characteristic function of $I_2 = (x_2-h_2,\ x_2+h_2)$, where I_1 and I_2 do not overlap (say $x_1 + h_1 \leqslant x_2 - h_2$, and $h_1 \geqslant h_2 > 0$), then

$$\hat{\chi}_1(x) = \frac{1}{\pi}\frac{\sin h_1 x}{x} e^{-ix_1 x}$$

$$\hat{\chi}_2(x) = \frac{1}{\pi}\frac{\sin h_2 x}{x} e^{-ix_2 x}$$

Hence,

$$\int_{-\infty}^{\infty} \hat{\chi}_1(x)\overline{\hat{\chi}_2(x)}\, dx = \frac{1}{\pi^2}\int_{-\infty}^{\infty} \frac{\sin h_1 x}{x}\,\frac{\sin h_2 x}{x}\, e^{i(x_2-x_1)x}\, dx.$$

If we set

$$f(x) = \frac{1}{\pi^2}\,\frac{\sin h_1 x}{x}\,\frac{\sin h_2 x}{x}$$

then $f(x)$ is an integrable function on E^1 and is equal (up to a constant factor) to the Fourier transform of the trapezoidal function $\chi_{h_1,h_2}(x)$ which is continuous and integrable (cf. Example 3 of §1). Since the Inversion Formula holds everywhere in this case, we have that, for some constant $c \neq 0$,

$$\int_{-\infty}^{\infty} \hat{\chi}_1(x)\overline{\hat{\chi}_2(x)}\, dx = \int_{-\infty}^{\infty} f(x) e^{i(x_2-x_1)x}\, dx = c\chi_{h_1,h_2}(x_2-x_1) = 0$$

because (as it is easily seen by a picture of I_1 and I_2 and by graphing the trapezoidal function χ_{h_1,h_2})

$$|x_2 - x_1| \geqslant h_1 + h_2$$

and therefore χ_{h_1,h_2} vanishes at the point $x_2 - x_1$.
Hence the Claim is proved.

Finally by Step 1 and using the fact the function $\chi_k(x)$ have non-overlapping supports, we obtain

$$\int_{-\infty}^{\infty} |\hat{f}(x)|^2\, dx = \sum_{k=1}^{m} |c_k|^2 \int_{-\infty}^{\infty} |\hat{\chi}_k(x)|^2\, dx = \frac{1}{2\pi} \sum_{k=1}^{m} |c_k|^2 \int_{-\infty}^{\infty} \chi_k^2(x)\, dx =$$

$$= \frac{1}{2\pi} \int_{-\infty}^{\infty} \Big| \sum_{k=1}^{m} c_k \chi_k(x) \Big|^2\, dx = \frac{1}{2\pi} \int_{-\infty}^{\infty} |f(x)|^2\, dx$$

whence, taking square roots, we see that Parseval's Formula holds.

Step 4: Simple functions on E^n.

This case can be handled exactly as the previous one. The details are left as an exercise to the reader.

Step 5: Extension by continuity.

We have seen that the Fourier transform $Tf = \hat{f}$ is defined on the class S of simple function, and satisfies Parseval's Formula

(*) $$\|\hat{f}\|_2 = (2\pi)^{-n/2} \|f\|_2 \qquad \text{for all } f \in S.$$

Hence T is a bounded (so,continuous) densely defined linear operator in L^2.

CLAIM: The operator T can be extended to the closure $\bar{S} = L^2$, with preservation of norm, that is, in such a way that formula (*) will hold for all $f \in L^2$.

In fact, if $f \in L^2$, choose a sequence $f_n \in S$ such that $f_n \to f$ in L^2, i.e. $\|f - f_n\|_2 \to 0$ as $n \to \infty$; in particular $\{f_n\}$ is a Cauchy sequence so

$$\|f_n - f_m\|_2 \to 0 \quad \text{as } m, n \to \infty, \text{ thus,}$$

by continuity of T,

$$\|Tf_n - Tf_m\|_2 \to 0 \quad \text{as } m, n \to \infty, \text{ i.e.}$$

$\{Tf_n\}$ is a Cauchy sequence. Therefore, by completeness of L^2, there is some $g \in L^2$ such that $Tf_n \to g$ in L^2, as $n \to \infty$. We define Tf by setting

$$Tf = g.$$

Tf is well-defined, i.e. is independent of the choice of $\{f_n\}$, since if $\{f_n'\}$ is another sequence in S such that

$$f_n' \to f \text{ in } L^2$$

and

$$Tf_n' \to g' \text{ in } L^2,$$

then the sequence $f_1, f_1', f_2, f_2', \ldots$ converges to f, and hence the sequence $Tf_1, Tf_1', Tf_2, Tf_2', \ldots$ also converges in L^2 therefore we must have that $g = g'$.

Finally, from the formula

$$\|Tf_m\|_2 = (2\pi)^{-n/2}\|f_m\|_2, \quad f_m \in S \text{ and } f_m \to f,$$

we obtain, passing to the limit (since the norm is a continuous function), that

$$\|\hat{f}\|_2 = \|Tf\|_2 = (2\pi)^{-n/2}\|f\|_2, \text{ for every } f \in L^2.$$

So the claim is proved.

This extension of the operator T is called extension by continuity.

Step 6: $Tf = \hat{f}$ is the limit, in L^2 norm, of the truncated Fourier transform $\hat{f}_R$.

We first prove that if $f \in L^2$ has bounded support then Tf is of the form

$$\hat{f}_R(x) = (2\pi)^{-n} \int_{|y|\leq R} f(y)e^{-i(x\cdot y)}\, dy.$$

Suppose that the support of f is contained in a sphere $|x| \leq R$. Choose a sequence of functions $f_k \in S$ with support in $|x| \leq R$ such that, as $k \to \infty$,

$$f_k \to f \text{ in } L^2.$$

The, by continuity of T,

$$Tf_k \to Tf \text{ in } L^2$$

and hence there exists a subsequence of $\{Tf_k\}$ which converges to Tf pointwise a.e. Moreover,

$$Tf_k(x) = (2\pi)^{-n} \int_{|y|\leq R} f_k(y)e^{-i(x\cdot y)}dy$$

and

$$\hat{f}_R(x) = (2\pi)^{-n} \int_{|y|\leq R} f(y)e^{-i(x\cdot y)}dy,$$

therefore

$$|Tf_k(x) - \hat{f}_R(x)| \leq (2\pi)^{-n} \int_{|y|\leq R} |f_k(y)-f(y)|\,dy \leq$$

(by Schwarz's Inequality)

$$\leq C\|f_k-f\|_2 \to 0, \text{ as } k \to \infty,$$

so the sequence $\{Tf_k\}$ converges pointwise to $\hat{f}_R$. Thus $Tf = \hat{f}_R$ a.e. .

Now for any $f \in L^2$, we consider the truncations

$$f_R(x) = \begin{cases} f(x) & \text{if } |x| \leq R \\ 0 & \text{if } |x| > R \end{cases}$$

As $R \to \infty$, $f_R \to f$ in L^2, so $Tf_R \to Tf$ in L^2. But, by the previous argument $Tf_R = \hat{f}_R$ a.e., therefore as $R \to \infty$

$$\hat{f}_R \to Tf = \hat{f} \text{ in } L^2.$$

Step 7: The Inversion Formula.

Finally, we must verify the Inversion Formula

$$(\dagger) \qquad f(x) = \int \hat{f}(y) e^{i(x \cdot y)} dy,$$

where the integral is understood to be the limit in L^2 norm of its truncations to sphere $|y| \leq R$.

It should be clear from our previous discussion that it is enough to prove ($\dagger$) for functions $f \in S$. Moreover, by linearity, it suffices to prove ($\dagger$) for characteristic functions of n-dimensional intervals, but this case reduces easily to the case of characteristic functions of intervals in E^1.

From the formula

$$\frac{1}{\pi} \int_{-\infty}^{\infty} \left(\frac{\sin x}{x} \right)^2 dx = 1$$

we obtain, intergrating by parts, that

$$\frac{1}{\pi} \int_{-\infty}^{\infty} \frac{\sin x}{x} = 1.$$

In fact,

$$\int_{-\infty}^{\infty} \frac{\sin^2 x}{x^2} dx = \frac{1}{2} \int_{-\infty}^{\infty} \frac{1-\cos 2x}{x^2} dx =$$

$$= \lim_{N \to \infty} \left\{ - \frac{1}{2} \frac{1-\cos 2x}{x} \Bigg|_{x=-N}^{x=N} + \int_{-N}^{N} \frac{\sin 2x}{x} dx \right\} =$$

$$= \int_{-\infty}^{\infty} \frac{\sin y}{y} dy.$$

Moreover, for any real number α

$$\text{(1)} \qquad \frac{1}{\pi}\int_{-\infty}^{\infty} \frac{\sin \alpha x}{x}\, dx = \operatorname{sgn} \alpha$$

where

$$\operatorname{sgn} \alpha = \begin{cases} \alpha/|\alpha| & \text{if } \alpha \neq 0 \\ 0 & \text{if } \alpha = 0 \end{cases}$$

Now, if $f(x) = \chi_{a,b}(x)$ is the characteristic function of the interval $(a,b) = (c-h,\ c+h)$, then

$$\hat{f}(x) = \frac{1}{\pi}\frac{\sin hx}{x} e^{-icx}, \qquad \text{and}$$

$$\int_{-\infty}^{\infty} \hat{f}(y) e^{ixy}\, dy = \frac{1}{\pi}\int_{-\infty}^{\infty} e^{i(x-c)y} \frac{\sin hy}{y}\, dy =$$

$$= \frac{1}{\pi}\int_{-\infty}^{\infty} \frac{\cos(x-c)y \sin hy}{y}\, dy =$$

$$= \frac{1}{2\pi}\int_{-\infty}^{\infty} \frac{\sin(x-c+h)y - \sin(x-c-h)y}{y}\, dy =$$

$$= \frac{1}{2\pi}\int_{-\infty}^{\infty} \frac{\sin(x-a)y}{y} - \frac{\sin(x-b)y}{y}\, dy =$$

by (1),

$$= \frac{1}{2}\left[\operatorname{sgn}(x-a) - \operatorname{sgn}(x-b)\right] = \begin{cases} 0 & \text{if } x \notin [a,b] \\ 1 & \text{if } x \in (a,b) \\ 1/2 & \text{if } x = a \text{ or } x = b. \end{cases}$$

Therefore,

$$\int_{-\infty}^{\infty} \hat{f}(y) e^{ixy}\, dy = f(x) \text{ a.e.}$$

This completes the proof of Theorem 9.

Q.E.D.

REMARK: If we set

$$T^*g(x) = \int g(y)e^{i(x\cdot y)}\,dy,$$

then the Inversion Formula (†) can be written in the form

$$f = T^*Tf$$

which shows that $T^* = T^{-1}$. We conclude from Th. 9 that the Fourier Transform T is a continuous linear isomorphism of L^2.

COROLLARY 1: (Plancherel's Formula) If $f,g \in L^2(E^n)$ then

$$\int \hat{f}(x)\overline{\hat{g}(x)}\,dx = (2\pi)^{-n}\int f(x)\,\overline{g(x)}\,dx.$$

Proof: Since $(f+g)\hat{} = \hat{f} + \hat{g}$ then, by Parseval's Formula,

$$\int|\hat{f} + \hat{g}|^2\,dx = (2\pi)^{-n}\int|f+g|^2\,dx.$$

Now,

$$|f+g|^2 = (f+g)(\bar{f}+\bar{g}) = |f|^2 + |g|^2 + f\bar{g} + \bar{f}g = |f|^2 + |g|^2 + 2\,\mathrm{Re}(f\bar{g})$$

and likewise

$$|\hat{f} + \hat{g}|^2 = |\hat{f}|^2 + |\hat{g}|^2 + 2\,\mathrm{Re}(\hat{f}\,\bar{\hat{g}}).$$

Hence, using Parseval's formula again, we obtain that

$$\mathrm{Re}\left\{\int \hat{f}\,\bar{\hat{g}}\,dx\right\} = (2\pi)^{-n}\,\mathrm{Re}\left\{\int f\bar{g}\,dx\right\}.$$

Replacing f by if, the above formula gives the corresponding equality for the imaginary parts of our integrals, and hence Plancherel's formula is proved.

Q.E.D.

We have deduced Plancherel's Formula from Parseval's Formula; conversely, taking g = f in the former one obtains the latter, so the two formulas are equivalent. They show that, up to a constant factor, the Fourier transform preserves the norm and the inner product of the Hilbert space L^2. In particular, Plancherel's Formula shows that the Fourier transform preserves orthogonality.

REMARK: If we define the Fourier transform T by the formula

$$Tf(x) = \hat{f}(x) = \int f(y)e^{-2\pi i(x\cdot y)}dy,$$

then
$$\|\hat{f}\|_2 = \|f\|_2 \quad \text{(Parseval's Formula)}$$

$$(\hat{f},\hat{g}) = (f,g) \quad \text{(Plancherel's Formula)}$$

and the Inversion Formula becomes

$$f(x) = \int \hat{f}(y)e^{2\pi i(x\cdot y)}\, dy = T^{*}Tf.$$

Therefore T is a unitary transformation of L^2.

EXERCISE: Let $f \in L^2(E^n)$. Prove that

a) $\hat{\bar{f}}(x) = \overline{\hat{f}(-x)}$,

b) $f(x) = (2\pi)^n\hat{\hat{f}}(-x)$.

Note that b) shows that every $f \in L^2$ is the Fourier transform of some other element of L^2.

The following result is the analogue of Th. 3 of §1.

Corollary 2: If f, $g \in L^2(E^n)$ then
$$\int \hat{f}(x)g(x)\, dx = \int f(x)\, \hat{g}(x)\, dx.$$

Proof: By the previous exercise it follows that if we set $\hat{f}(x) = u(x)$ and $g(x) = \overline{v(x)}$, then taking Fourier transform we obtain

$$(2\pi)^n\hat{u}(x) = f(-x)$$

and

$$\overline{\hat{v}(x)} = \hat{\bar{v}}(-x) = \hat{g}(-x).$$

Hence, using Plancherel's Formula,

$$\int \hat{f}(x)\, g(x)\, dx = \int u(x)\, \overline{v(x)}\, dx = (2\ )^n \int \hat{u}(x)\, \overline{\hat{v}(x)}\, dx =$$

$$= \int f(-x)\hat{g}(-x)\, dx = \int f(x)\, \hat{g}(x)\, dx.$$

Q.E.D.

We recall from §1 that if $f, g \in L(E^n)$ and we set

$$h(x) = (f*g)(x) = (2\pi)^{-n} \int f(y)g(x-y)\, dy$$

then $h \in L$ also, and $\hat{h} = \hat{f}\hat{g}$. On the other hand, if $f, g \in L^2$ then $h = f*g$ is merely a bounded function (by Schwarz's Inequality) so, in general its Fourier transform is not defined. The following theorem, however, provides a useful substitute result.

THEOREM 10: If $f \in L^2(E^n)$ and $g \in L(E^n)$, then $h = f*g$ belongs to $L^2(E^n)$ and $\hat{h} = \hat{f}\,\hat{g}$.

Proof: The first part of the conclusion follows from Young's Theorem (Th. 2 of Chapter I) in the case $p = 2$. It remains to show that $\hat{h} = \hat{f}\,\hat{g}$.

Consider the truncations

$$f_k(x) = \begin{cases} f(x) & \text{if } |x| \leq k \\ 0 & \text{if } |x| > k. \end{cases}$$

We know that $f_k \to f$ in L^2, i.e.

(1) $$\|f - f_k\|_2 \to 0 \quad \text{as } k \to \infty.$$

Hence, by continuity of the Fourier transform (Parseval's Formula), we have that $\hat{f}_k \to \hat{f}$ in L^2, i.e.

(2) $$\|\hat{f}_k - \hat{f}\|_2 \to 0 \quad \text{as } k \to \infty.$$

Moreover, by Schwarz's Inequality, the functions f_k belong to L, hence by Th. 1 of Chapter I the convolutions $h_k = f_k*g$ also belong to L, and by Th. 2 of §1, we have that

(3) $\hat{h}_k = \hat{f}_k\hat{g}$.

Since $g \in L$ and $f_k \in L^2$ then by Young's Theorem $h_k \in L^2$. Moreover

$h_k \to h$ in L^2, because

$$h - h_k = f*g - f_k*g = (f-f_k)*g$$

so by Young's Theorem and (1) we have that

$$\|h - h_k\|_2 \leq \|f - f_k\|_2 \|g\|_1 \to 0 \quad \text{as } k \to \infty.$$

Hence $\hat{h}_k \to \hat{h}$ in L^2, that is $\|\hat{h}_k - \hat{h}\|_2 \to 0$ as $k \to \infty$.

On the other hand, we have that $\hat{h}_k \to \hat{f}\hat{g}$ in L^2. In fact, by (3),

$$\hat{h}_k = \hat{f}_k\hat{g} = (\hat{f}_k - \hat{f})\hat{g} + \hat{f}\hat{g}$$

so

$$\|\hat{h}_k - \hat{f}\hat{g}\|_2 = \|(\hat{f}_k - \hat{f})\hat{g}\|_2 \leq C\|\hat{f}_k - \hat{f}\|_2 \to 0 \text{ as } k \to \infty$$

by (2) and the fact that $\hat{g}$ is bounded since $g \in L$.

Therefore we conclude that $\hat{h} = \hat{f}\hat{g}$.

Q.E.D.

REMARK: If $h = f*g$ where $f, g \in L$, then $\hat{h}(x) = \hat{f}(x)\hat{g}(x)$ for every $x \in E^n$. On the other hand, in the preceding theorem $\hat{h} = \hat{f}\hat{g}$ as elements of L^2, that is they coincide almost everywhere.

We conclude this chapter with some comments about the Fourier transform in the spaces $L^p(E^n)$.

We have seen that the Fourier transform $Tf = \hat{f}$, where

$$\hat{f}(x) = (2\pi)^{-n} \int_{E^n} f(y) e^{-i(x \cdot y)} dy$$

is defined in L^1 and in L^2, and is a bounded operator from L^1 into L^∞ and from L^2 onto L^2. More precisely;

(i) $\qquad \|\hat{f}\|_\infty \leq (2\pi)^{-n} \|f\|_1, \qquad$ for all $f \in L^1(E^n)$, and

(ii) $\qquad \|\hat{f}\|_2 = (2\pi)^{-n/2} \|f\|_2, \qquad$ for all $f \in L^2(E^n)$.

Suppose now that $f \in L^p$, where $1 < p < 2$, and split f into two parts, i.e. $f = f_1 + f_2$ where

$$f_1(x) = \begin{cases} f(x) & \text{whenever } |f(x)| \leqslant 1 \\ 0 & \text{otherwise} \end{cases}$$

and $f_2 = f - f_1$. Clearly f_1 and f_2 also belongs to L^p. Moreover, by definition, $|f_1(x)| \leqslant 1$, so $|f_1(x)|^2 \leqslant |f_1(x)|^p$ since $1 < p < 2$; hence $f_1 \in L^2$. Similarly we see that $f_2 \in L$. Therefore $\hat{f}_1(x)$ exists a.e. and $\hat{f}_2(x)$ exists everywhere, and we can define the Fourier transform f by the formula

$$\hat{f} = \hat{f}_1 + \hat{f}_2.$$

This shows that the Fourier transform can be defined also in the spaces $L^p(E^n)$ with $1 < p < 2$. Also, using Hölder's Inequality, we verify the existence of the truncated integrals

$$\hat{f}_R(x) = (2\pi)^{-n} \int_{|y| \leqslant R} f(y) e^{-i(x \cdot y)} dy.$$

THEOREM of Hausdorff-Young: Let $f \in L^p(E^n)$, $1 < p \leqslant 2$, and let $q = p/(p-1)$ be the conjugate exponent of p (thus $2 \leqslant q < \infty$). Then the Fourier transform

$$\hat{f}(x) = (2\pi)^{-n} \int f(y) e^{-i(x \cdot y)} dy$$

exists as a limit in L^q norm of the $\hat{f}_R(x)$, $R \to \infty$, and

$$\|\hat{f}\|_q \leqslant (2\pi)^{-n/p} \|f\|_p.$$

Conversely, any $f \in L^p$, $1 < p \leqslant 2$, is the Fourier transform of some $g \in L^q$, i.e.

$$f(x) = \underset{R \to \infty}{\text{limit in } L^p}\ (2\pi)^{-n} \int_{|y| \leqslant R} g(y) e^{-i(x \cdot y)} dy,$$

and

$$\|g\|_q \leqslant (2\pi)^{-n/p} \|f\|_p.$$

The proof of this theorem can be deduced from the formulas (i) and (ii) above using the theorem of M. Riesz-Thorin on the interpolation of linear operators (Cf, Zygmund: Trigonometric Series, Vol. II.)

REMARK: Functions in L^p, with $p > 2$, in general have no Fourier transform. However, since such functions are locally integrable and small at ∞, they can be viewed as "tempered distributions" and one can define their Fourier transforms in the sense of the theory of distributions (Cf. L. Schwartz: Thèorie des Distributions, Vol. I and II.)

EXERCISES

1. Prove that there exists no function $g \in L^1$ such that $g*f = f$ for all $f \in L^1$. (Hint: assume the contrary and use the Theorem of Riemann-Lebesgue).
2. Prove that if $\{f_k\}$ is an orthonormal system in $L^2(E^n)$, then so is $\{g_k\}$ where $g_k = (2\pi)^{-n/2}\hat{f}_k$; moreover if the first system is complete so is the second.
3. Let $\{n_k\}$ be a sequence of real numbers such that $|n_k - n_l| > 2h$ for all k and l. Show that the system $e^{in_k x}$ is orthogonal over E^1 with respect to the weight function $\dfrac{\sin^2 hx}{x^2}$.

   (Hint: Let f_k be the characteristic function of the interval $(n_k - h, n_k + h)$ and note that all these intervals are disjoint; then use the previous exercise).
4. Show that if the sequence $\{\hat{f}_k\}$ has a limit in the metric L^2, then so does the sequence $\{f_k\}$. If $\hat{f}_k \to g$, then $f_k \to \check{g}$, the Inverse Fourier transform of g.
5. Prove that if $f \in L^2(E^1)$ then the integral $\displaystyle\int_{-\infty}^{\infty} \hat{f}(y)e^{ixy}dy$ is

summable to f(x) almost everywhere, and in L^2 norm, by the Abel-Poisson method.

CHAPTER III

THE HILBERT TRANSFORM

Suppose that $f \in L^p(E^1)$, $1 \leq p < \infty$, and consider the principal value convolution

$$\tilde{f}(x) = \text{p.v. } \frac{1}{\pi} \int_{-\infty}^{\infty} \frac{f(t)}{x-t}\, dt = \lim_{\varepsilon \to 0^+} \frac{1}{\pi} \int_{|x-t|>\varepsilon} \frac{f(t)}{x-t}\, dt$$

which is called the Hilbert Transform of f. Our main objectives in this chapter will be to establish the existence of $\tilde{f}$ and to prove the classical result of M. Riesz which states that the operator $Hf = \tilde{f}$ is continuous in L^p for $1 < p < \infty$.

Formally, $\tilde{f}$ is the convolution of f with the kernel $1/x$ but this convolution must be taken in the principal value sense since our kernel is not integrable on E^1. If we write

$$\tilde{f}(x) = \lim_{\varepsilon \to 0} \tilde{f}_\varepsilon(x)$$

where

$$\tilde{f}_\varepsilon(x) = \frac{1}{\pi} \int_{|x-t|>\varepsilon} \frac{f(t)}{x-t}\, dt$$

and $f \in L^p$, $1 \leq p < \infty$, we see, by Hölder's Inequality, that $\tilde{f}_\varepsilon$ exists, since outside an ε-neighborhood of the origin the kernel $1/x$ belongs to L^q for every $q > 1$. If $f \in L \cap L^p$, $p > 1$, $\tilde{f}_\varepsilon \in L^p$ by Th. 2 of Chapter I.

On the other hand,

$$\tilde{f}(x) = \lim_{\varepsilon \to 0} \frac{1}{\pi} \int_{|x-t|>\varepsilon} \frac{f(t)}{x-t}\, dt = \text{(letting } u = x-t)$$

$$= \lim_{\varepsilon \to 0} \frac{1}{\pi} \int_{|u|>\varepsilon} \frac{f(x-u)}{u}\, du = \text{(replacing } u \text{ by } -u)$$

$$= \lim_{\varepsilon\to 0} \frac{-1}{\pi} \int_{|u|>\varepsilon} \frac{f(x+u)}{u} du = \lim_{\varepsilon\to 0} \frac{-1}{\pi} \int_{\varepsilon}^{\infty} \frac{f(x+u)-f(x-u)}{u} du =$$

$$= \frac{-1}{\pi} \int_{0}^{\infty} \frac{f(x+u)-f(x-u)}{u} du$$

and this last integral exists if f satisfies, in addition, a mild smoothness condition. Namely, if f satisfies a Hölder condition of order α, $0 < \alpha \leq 1$, that is

(i) $$|f(x)-f(y)| \leq C|x-y|^{\alpha} \quad \text{if } |x-y| \leq \delta,$$

then

$$|\tilde{f}(x)| \leq \int_{0}^{\infty} \frac{|f(x+u)-f(x-u)|}{|u|} du = \int_{0}^{\delta} + \int_{\delta}^{\infty} = I_1 + I_2$$

and, by (i)

$$I_1 \leq \int_{0}^{\delta} C|u|^{\alpha-1} du < \infty, \text{ since } \alpha > 0,$$

while I_2 is finite, by Hölder's Inequality, for $f \in L^p$, $1 \leq p < \infty$.

We shall examine first the L^2-theory of the Hilbert transform.

THEOREM 1: Let $f \in L^2(E^1)$, $0 < \varepsilon < \omega < \infty$, and set

$$\tilde{f}_{\varepsilon,\omega}(x) = \frac{1}{\pi} \int_{\varepsilon<|x-t|<\omega} \frac{f(t)}{x-t} dt.$$

Then,

1) $\|\tilde{f}_{\varepsilon,\omega}\|_2 \leq A\|f\|_2$, for some absolute constant $A > 0$.

2) There exists a function $\tilde{f} \in L^2$ such that $\|\tilde{f}_{\varepsilon,\omega}-\tilde{f}\|_2 \to 0$, as $\varepsilon\to 0$ and $\omega \to \infty$ simultaneously or successively.

3) $\|\tilde{f}\|_2 = \|f\|_2$.

4) $\tilde{\tilde{f}} = -f$.

5) $\tilde{f}(x) = \lim_{\varepsilon\to 0} \tilde{f}_{\varepsilon}(x)$ pointwise almost everywhere.

Proof: If we let

$$K_{\varepsilon,\omega}(x) = \begin{cases} 2/x & \text{if } \varepsilon < |x| < \omega \\ 0 & \text{elsewhere} \end{cases}$$

then

$$\tilde{f}_{\varepsilon,\omega}(x) = \frac{1}{2\pi}\int_{-\infty}^{\infty} f(t)K_{\varepsilon,\omega}(x-t)\,dt = (f*K_{\varepsilon,\omega})(x)$$

is the (normalized) convolution of $f \in L^2$ with the kernel $K_{\varepsilon,\omega} \in L$. Hence, by Th. 10 of last chapter, $\tilde{f}_{\varepsilon,\omega}$ belongs to L^2 and

(a) $$\hat{\tilde{f}}_{\varepsilon,\omega}(x) = \hat{f}(x)\hat{K}_{\varepsilon,\omega}(x) \text{ a.e.}$$

Now,

$$\hat{K}_{\varepsilon,\omega}(x) - \frac{1}{2\pi}\int_{\varepsilon<|t|<\omega} \frac{2}{t}e^{-ixt}dt = \frac{1}{\pi}\int_{\varepsilon}^{\omega}\frac{e^{-ixt}-e^{ixt}}{t}\,dt =$$

$$= -\frac{2i}{\pi}\int_{\varepsilon}^{\omega}\frac{\sin xt}{t}\,dt = -\frac{2i}{\pi}(\operatorname{sgn} x)\int_{\varepsilon|x|}^{\omega|x|}\frac{\sin u}{u}\,du$$

where the last integral is uniformly bounded and converges to

$$\int_0^{\infty}\frac{\sin u}{u}\,du = \frac{\pi}{2}$$

as $\varepsilon \to 0$ and $\omega \to \infty$ simultaneously or successively. Hence we have that, for every x, ε and ω,

(b) $|\hat{K}_{\varepsilon,\omega}(x)| \leq A$

(c) $\hat{K}_{\varepsilon,\omega}(x) \to -i \operatorname{sgn} x$ as $\varepsilon \to 0$ and $\omega \to \infty$.

From (a) and (b) it follows that $\|\hat{\tilde{f}}_{\varepsilon,\omega}\|_2 \leq A\|\hat{f}\|_2$, whence, by Parseval's Formula

1) $\|\tilde{f}_{\varepsilon,\omega}\|_2 \leq A\|f\|_2$.

From (a) (b) and (c) it follows that $\hat{\tilde{f}}_{\varepsilon,\omega} \to (-i \operatorname{sgn} x)\hat{f}$ in L^2

norm, since

$$\|\hat{\tilde{f}}_{\varepsilon,\omega}(x) + (i \operatorname{sgn} x)\hat{f}(x)\|_2 = \|[\hat{K}_{\varepsilon,\omega}(x)+i \operatorname{sgn} x]\hat{f}(x)\|_2 \to 0$$

as $\varepsilon \to 0$ and $\omega \to \infty$, by Lebesgue's Theorem of dominated convergence. Hence, taking Inverse Fourier transform, (Cf. Exercise 4 at the end of last chapter) we have that

$$\tilde{f}_{\varepsilon,\omega} \to \tilde{f} \text{ in } L^2 \text{ norm,}$$

for some function $\tilde{f} \in L^2$ such that

(d) $$\hat{\tilde{f}}(x) = (-i \operatorname{sgn} x)\hat{f}(x).$$

This proves conclusion 2).

Moreover, from (d), we see that $\|\hat{\tilde{f}}\|_2 = \|\hat{f}\|_2$, so, by Parseval's

3) $\|\tilde{f}\|_2 = \|f\|_2$.

Using (d) again we have that

$$\hat{\tilde{\tilde{f}}}(x) = (-i \operatorname{sgn} x)\hat{\tilde{f}}(x) = (-i \operatorname{sgn} x)^2 \hat{f}(x) = -\hat{f}(x), \quad \text{so}$$

4) $$\tilde{\tilde{f}} = - f.$$

Conclusions 1) and 2) are also valid for $\tilde{f}_\varepsilon$ as will be seen later, and more generally, in Theorem 3.

It remains to be proved that $\tilde{f}_\varepsilon(x)$ also converges to $\tilde{f}(x)$ pointwise almost everywhere.

By the Inversion Formula for Fourier transforms, and by (d), it follows that

$$\tilde{f}(x) = \int_{-\infty}^{\infty} \hat{\tilde{f}}(t)e^{ixt}\, dt = \int_{-\infty}^{\infty} \hat{f}(t)(-i \operatorname{sgn} t)\, e^{ixt}\, dt$$

where the integral converges in L^2 norm as a limit of truncated integrals, and moreover (cf. Exercise 5 at the end of last chapter) it is summable a.e. by the Abel-Poisson method, for example. So, for almost every x,

$$\tilde{f}(x) = \lim_{\varepsilon \to 0^+} \int_{-\infty}^{\infty} \hat{f}(t)(-i \operatorname{sgn} t) e^{ixt} e^{-\varepsilon|t|} \, dt =$$

$$= \lim_{\varepsilon \to 0^+} \int_{-\infty}^{\infty} \widehat{f(x+t)} \{(-i \operatorname{sgn} t) e^{-\varepsilon|t|}\} \, dt.$$

Since sgn t = sgnε t, for all $\varepsilon > 0$, and the function

$$H(t) = (-i \operatorname{sgn} t) e^{-|t|}$$

belongs to L^2, so, by Coro. 2 of Th. 9 of last chapter, we have that for almost every x

$$\tilde{f}(x) = \lim_{\varepsilon \to 0^+} \int_{-\infty}^{\infty} \widehat{f(x+t)} H(\varepsilon t) \, dt =$$

$$= \lim_{\varepsilon \to 0^+} \int_{-\infty}^{\infty} f(x+t) \widehat{H(\varepsilon t)} \, dt.$$

We shall calculate next $\widehat{H(\varepsilon t)} = \frac{1}{\varepsilon} \hat{H}(\frac{t}{\varepsilon})$. Recall that

$$(e^{-|t|})^{\wedge}(x) = \frac{1}{\pi} \frac{1}{1+x^2}$$

so

$$(e^{-\varepsilon|t|})^{\wedge}(x) = \frac{1}{\pi} \frac{\varepsilon}{\varepsilon^2+x^2}$$

which is the Poisson Kernel in E^1. Now,

$$[(\operatorname{sgn} t) e^{-|t|}]^{\wedge}(x) = \frac{1}{2\pi} \int_0^{\infty} e^{-t} e^{-ixt} dt - \frac{1}{2\pi} \int_{-\infty}^{0} e^{+t} e^{-ixt} dt =$$

$$= \frac{1}{2\pi} \int_0^{\infty} e^{-t(1+ix)} dt - \frac{1}{2\pi} \int_0^{\infty} e^{-t(1-ix)} dt =$$

$$= \frac{1}{2\pi} \left[\frac{1}{1+ix} - \frac{1}{1-ix} \right] = \frac{-i}{\pi} \frac{x}{1+x^2}$$

which is not an integrable function but belongs to L^p for every $p > 1$, so, in particular, is in L^2. Since $H(t) = (-i \operatorname{sgn} t)e^{-|t|}$, we have that

$$\hat{H}(x) = \frac{-1}{\pi} \frac{x}{1+x^2}$$

$$\widehat{H(\varepsilon x)} = \frac{-1}{\pi} \frac{x}{\varepsilon^2+x^2}$$

which is, up to sign, the <u>Conjugate Poisson Kernel</u> in E^1.

Therefore we have obtained that almost everywhere

$$\tilde{f}(x) = \lim_{\varepsilon \to 0^+} \left[\frac{-1}{\pi} \int_{-\infty}^{\infty} f(x+t) \frac{t}{\varepsilon^2+t^2} dt \right] =$$

$$= \lim_{\varepsilon \to 0^+} \left[\frac{1}{\pi} \int_{-\infty}^{\infty} f(x-t) \frac{t}{\varepsilon^2+t^2} dt \right]$$

where the expression in brackets, which is the convolution of f with the Conjugate Poisson Kernel, is called the <u>Conjugate Poisson Integral</u> of f. In other words, the Conjugate Poisson Integral of f converges a.e. to $\tilde{f}(x)$. Hence, in order to prove that the truncated Hilbert transform

$$\tilde{f}_\varepsilon(x) = \frac{1}{\pi} \int_{|x-t|>\varepsilon} \frac{f(t)}{x-t} dt = \frac{1}{\pi} \int_{|t|>\varepsilon} \frac{f(x-t)}{t} dt$$

also converges a.e. to $\tilde{f}(x)$, it suffices to show that it is equi-convergent with the Conjugate Poisson Integral of f, that is

$$\text{(e)} \quad \int_{-\infty}^{\infty} f(x-t) \frac{t}{\varepsilon^2+t^2} dt - \int_{|t|>\varepsilon} f(x-t) \frac{1}{t} dt \to 0$$

almost everywhere as $\varepsilon \to 0$.

Now, the expression in (e) is the convolution of f with the Kernel

$$K_\varepsilon(t) = \begin{cases} \dfrac{t}{\varepsilon^2+t^2} & \text{if } |t| \leq \varepsilon \\[2ex] \dfrac{t}{\varepsilon^2+t^2} - \dfrac{1}{t} & \text{if } |t| > \varepsilon \end{cases}$$

and $K_\varepsilon(t) = \frac{1}{\varepsilon} K\left(\frac{t}{\varepsilon}\right)$, where

$$K(t) = \begin{cases} \dfrac{t}{1+t^2} & \text{if } |t| \leq 1 \\[2ex] \dfrac{t}{1+t^2} - \dfrac{1}{t} & \text{if } |t| > 1. \end{cases}$$

Although it is the difference of two non-integrable kernels, $K(t)$ is integrable; in fact we have that $K(t)$ is bounded and, for $|t| > 1$,

$$K(t) = \frac{-1}{t+t^3} = \mathcal{O}\left(\frac{1}{t^3}\right)$$

Moreover,

$$\int_{-\infty}^{\infty} K(t)\, dt = 0$$

since the function $K(t)$ is odd. In other words $K(t)$ satisfies the hypotheses of Th. 6 of Chapter I with the condition $\int K(t)\, dt = 1$ replaced by the condition $\int K(t)\, dt = 0$. Therefore, we conclude that, for almost every x,

$$\lim_{\varepsilon \to 0} \int_{-\infty}^{\infty} f(x-t) K_\varepsilon(t)\, dt = f(x) \int_{-\infty}^{\infty} K(t)\, dt = 0 \text{ which proves (e).}$$

Hence,

5) $\qquad \tilde{f}(x) = \lim_{\varepsilon \to 0} \tilde{f}_\varepsilon(x) \qquad$ pointwise a.e.

and the proof of our theorem is complete.

Q.E.D.

We have shown that if $f \in L^2(E^1)$, then the Hilbert transform

$$\tilde{f}(x) = \text{p.v.} \frac{1}{\pi} \int_{-\infty}^{\infty} \frac{f(t)}{x-t}\, dt = \lim_{\varepsilon \to 0} \frac{1}{\pi} \int_{|x-t|>\varepsilon} \frac{f(t)}{x-t}\, dt = \lim_{\varepsilon \to 0} \tilde{f}_{\varepsilon}(x)$$

exists pointwise a.e. and also as a limit in L^2 norm. Moreover, from

3) $\qquad \|\tilde{f}\|_2 = \|f\|_2$

we see that the linear operator $Hf = \tilde{f}$ is an isometry of L^2, and from

4) $\qquad \tilde{\tilde{f}} = -f$

we see that $H^2 = -I$, where I denotes the identity operator, i.e. the inverse operator $H^{-1} = -H$, and we have the inversion formula

$$f(x) = \text{p.v.} \frac{-1}{\pi} \int_{-\infty}^{\infty} \frac{\tilde{f}(t)}{x-t}\, dt.$$

REMARK: In view of the formula

(d) $\qquad \hat{\tilde{f}}(x) = (-i \operatorname{sgn} x)\hat{f}(x)$

if we let $\sigma(H)(x) = -i \operatorname{sgn} x$ and denote by F and F^{-1} the Fourier and the Inverse Fourier transforms, we see that we can represent the operator H in the form

$$H = F^{-1}\sigma(H)F.$$

Therefore, up to isomorphisms of L^2, the operator H corresponds to multiplication by the function $\sigma(H)$ which is called the <u>symbol</u> of the operator H. We note that formulas 3) and 4), above, are immediate consequences of the facts that $|\sigma(H)| = 1$ and that $\sigma(H)^2 = -1$, respectively. The function $\sigma(H)$ is the principal value Fourier transform of the kernel of H.

The following simple properties of the Hilbert transform $Hf = \tilde{f}$, are direct consequences of the definition.

I. $\quad \widetilde{f + ig} = \tilde{f} + i\,\tilde{g}$.

In particular, if f is real valued so is $\tilde{f}$. Moreover,

$\tilde{\bar{f}} = \bar{\tilde{f}}$, i.e. H commutes with conjugation.

II. $\widetilde{f(x+a)} = \tilde{f}(x+a)$, i.e. H commutes with translations.

III. If $\alpha \neq 0$, $\widetilde{f(\alpha x)} = (\operatorname{sgn} \alpha)\tilde{f}(\alpha x)$.

In particular, H commutes with multiplication by positive scalars (homotheties).

Corollary 1: If f, g $\in L^2(E^1)$, then

(a) $\int f \bar{g}\, dx = \int \tilde{f}\, \bar{\tilde{g}}\, dx$

(b) $\int f g\, dx = \int \tilde{f}\, \tilde{g}\, dx$

(c) $\int \tilde{f} g\, dx = -\int f \tilde{g}\, dx$

(d) The Hilbert transform is a unitary operator in L^2.

Proof: (a) follows from the formula $\|\tilde{f}\|_2 = \|f\|_2$, by the same argument that was used to deduce Plancherel's formula from Parseval's formula. In the inner product notation it becomes

(a') $(f,g) = (\tilde{f},\tilde{g})$

and shows that H preserves the inner product in L^2.

(b) follows directly from (a) if we replace $\bar{g}$ by g and use the fact that $\bar{\tilde{g}} = \tilde{\bar{g}}$.

To prove (c), we note that using (b) and the fact that $\tilde{\tilde{f}} = -f$ we obtain

$$\int \tilde{f} g\, dx = \int \tilde{\tilde{f}}\, \tilde{g}\, dx = -\int f \tilde{g}\, dx.$$

Finally, since $\tilde{\bar{g}} = \bar{\tilde{g}}$, replacing g by $\bar{g}$ we can write (c) in the form

(c') $(\tilde{f},g) = -(f,\tilde{g})$

which shows that the adjoint operator H* is equal to -H. On the other hand $-H = H^{-1}$, so we have that $H^* = H^{-1}$; in other words, the Hilbert transform is a unitary operator in L^2.

Q.E.D.

EXAMPLES:

1. Let $f(x) = \chi_{a,b}(x)$ be the characteristic function of the finite interval $I = (a,b)$. Then

$$\tilde{f}(x) = \frac{1}{\pi} \log \left| \frac{a-x}{b-x} \right| .$$

In fact, if $x \notin [a,b]$,

$$\tilde{f}(x) = \frac{1}{\pi} \int_a^b \frac{dt}{x-t} = \frac{-1}{\pi} \int_a^b \frac{dt}{t-x} = \frac{1}{\pi} \log \left| \frac{a-x}{b-x} \right| .$$

If $x \in (a,b)$ and $\varepsilon > 0$ is sufficiently small, then

$$\tilde{f}_\varepsilon(x) = \frac{1}{\pi} \left\{ \int_a^{x-\varepsilon} \frac{dt}{x-t} + \int_{x+\varepsilon}^{b} \frac{dt}{x-t} \right\} = \frac{1}{\pi} \log \left| \frac{a-x}{b-x} \right|$$

and letting $\varepsilon \to 0$ the result follows.

We note that as $|x| \to \infty$

$$\log \left| \frac{x-a}{x-b} \right| = \log \left| 1 + \frac{b-a}{x-b} \right| \sim \frac{|I|}{|x-b|} \sim \frac{|I|}{|x|}$$

so $\tilde{f}(x)$ is not integrable. Therefore the Hilbert transform does not preserve the class L^1.

2. Let $P(x) = \frac{1}{\pi} \frac{1}{1+x^2}$, $x \in E^1$. Then

$$\tilde{P}(x) = \frac{1}{\pi} \frac{x}{1+x^2} .$$

We know that $P(x) = \hat{f}(x)$ where $f(x) = e^{-|x|}$. Since $f \in L^2(E^1)$ then $f(x) = 2\pi \hat{\hat{f}}(-x)$; we have that

$$\hat{P}(x) = \hat{\hat{f}}(x) = \frac{1}{2\pi} f(-x) = \frac{1}{2\pi} e^{-|x|} .$$

Hence,

$$\hat{\tilde{P}}(x) = \hat{P}(x)\ (-i \operatorname{sgn} x) = \frac{1}{2\pi} e^{-|x|}\ (-i \operatorname{sgn} x)$$

so

$$\hat{\hat{\tilde{P}}}(x) = 2\pi\, P\,(-x) = [(-i \operatorname{sgn} t)e^{-|t|}]\hat{}\,(-x) = \frac{1}{\pi}\frac{x}{1+x^2}$$

as we calculated in the proof of Th. 1.

3. Let $P(x) = \frac{1}{\pi}\ \frac{1}{1+x^2}$ as before. Then, for all $\varepsilon > 0$,

$$P_\varepsilon(x) = \frac{1}{\varepsilon}\, P\,(\frac{x}{\varepsilon}) = \frac{1}{\pi}\ \frac{\varepsilon}{\varepsilon^2+x^2}$$

is the Poisson Kernel, which is a positive integrable kernel such that

$$\|P_\varepsilon\|_1 = \int_{-\infty}^{\infty} P_\varepsilon(x)\, dx = 1.$$

By property III above and the previous example, its Hilbert transform

$$\tilde{P}_\varepsilon(x) = \frac{1}{\varepsilon}\,\tilde{P}\,(\frac{x}{\varepsilon}) = \frac{1}{\pi}\,\frac{x}{\varepsilon^2+x^2}$$

is the Conjugate Poisson Kernel, which is not integrable but belongs to L^p for all $p > 1$.

If $f \in L^2$, then $\tilde{f} \in L^2$ and the <u>Poisson Integral</u> of $\tilde{f}$, i.e. the convolution of f with the Poisson Kernel, exists a.e. and belongs again to L^2, by Young's Theorem. Moreover, since $P_\varepsilon(x)$ is even,

$$\int_{-\infty}^{\infty} \tilde{f}(x-t)P_\varepsilon(t)\, dt = \int_{-\infty}^{\infty} \tilde{f}(x+t)P_\varepsilon(t)\, dt =$$

$$= \int_{-\infty}^{\infty} \widetilde{f(x+t)}P_\varepsilon(t)\, dt = -\int_{-\infty}^{\infty} f(x+t)\widetilde{P_\varepsilon}(t)\, dt$$

by (c) of Coro. 1, and since $\tilde{P}_\varepsilon(t)$ is odd we obtain that

$$(*)\qquad \int_{-\infty}^{\infty} f(x-t)P_\varepsilon(t)\, dt = \int_{-\infty}^{\infty} f(x-t)\widetilde{P_\varepsilon}(t)\, dt.$$

In other words, the Poisson Integral of $\tilde{f}$ is equal to the Conjugate Poisson Integral of f.

We now pass to the study of the Hilbert transform in the spaces L^p, where $1 < p < \infty$.

Suppose that $f \in L^p(E^1)$, $1 < p < \infty$, and consider the function

$$F(z) = \frac{1}{\pi i} \int_{-\infty}^{\infty} \frac{f(t)}{t-z}\, dt, \quad Z = x + iy,$$

in the upper half plane $y = \mathrm{Im}(z) > 0$. Since $(t-z)^{-1}$, as a function of t, belongs to $L^q(E^1)$ for every $q > 1$, the integral is finite by Hölder's Inequality and so F(z) is defined.

Moreover, F(z) is holomorphic in the upper half plane. In fact, if f(t) has finite support then

$$F(z) = \frac{1}{\pi i} \int_{-a}^{a} \frac{f(t)}{t-z}\, dt$$

and

$$\frac{1}{\Delta z}\,[F(z+\Delta z)-F(z)] = \frac{1}{\pi i} \int_{-a}^{a} f(t)\, \frac{1}{\Delta z} \left[\frac{1}{t-z-\Delta z} - \frac{1}{t-z} \right] dt =$$

$$= \frac{1}{\pi i} \int_{-a}^{a} f(t) \left[\frac{1}{(t-z-\Delta z)(t-z)} \right] dt.$$

But, as $\Delta z \to 0$, the expression in brackets converges uniformly to $(t-z)^{-2}$, for $-a \leq t \leq a$, hence

$$F'(z) = \frac{1}{\pi i} \int_{-a}^{a} \frac{f(t)}{(t-z)^2}\, dt$$

exists and F(z) is holomorphic in $\mathrm{Im}(z) > 0$. For a general $f \in L^p$, we consider the truncations

$$f_n(t) = \begin{cases} f(t) & \text{if } |t| \leq n \\ 0 & \text{otherwise;} \end{cases}$$

then the functions

$$F_n(z) = \frac{1}{\pi i} \int_{-\infty}^{\infty} \frac{f_n(t)}{t-z}\, dt$$

are all holomorphic and converge uniformly to $F(z)$, near each z in $\operatorname{Im}(z) > 0$, as $n \to \infty$. Therefore $F(z)$ is holomorphic in $\operatorname{Im}(z) > 0$.

Now, with $z = x+iy$ and $y > 0$, we have that

$$\frac{1}{t-z} = \frac{1}{t-x-iy} = \frac{t-x+iy}{(t-x)^2+y^2} = \frac{iy}{(t-x)^2+y^2} - \frac{x-t}{(t-x)^2+y^2}$$

so,

$$\frac{1}{\pi i}\ \frac{1}{t-z} = \frac{1}{\pi}\ \frac{y}{(x-t)^2+y^2} + i\,\frac{1}{\pi}\ \frac{x-t}{(x-t)^2+y^2} = P_y(x-t) + i\,\tilde{P}_y(x-t)$$

where $P_y(x)$ is the Poisson Kernel (with ε replaced by $y > 0$) and $\tilde{P}_y(x)$ is the Conjugate Poisson Kernel. Therefore,

$$F(z) = F(x+iy) = \int_{-\infty}^{\infty} f(t)P_y(x-t)\,dt + i \int_{-\infty}^{\infty} f(t)\,\tilde{P}_y(x-t)\,dt =$$

$$= U\ (x,y) + iV(x,y)$$

where $U(x,y)$ is the Poisson Integral of f and $V(x,y)$ is the Conjugate Poisson Integral of f.

In particular, if f is real valued, since $F(z)$ is holomorphic in $y > 0$ it follows that U and V are conjugate harmonic functions, in $y > 0$. Moreover, we know that for all $f \in L^p(E^1)$, with $1 \leqslant p < \infty$,

$$U(x,y) = (f*P_y)(x) \to f(x) \text{ a.e., as } y \to 0^+,$$

therefore the Poisson Integral $U(x,y)$ is a solution of the Dirichlet Problem for the upper half plane, with boundary condition $f(x)$ on the real line $y = 0$.

We have also proved that if $f \in L^2(E^1)$, then

$$V(x,y) = (f*\tilde{P}_y)(x) = (\tilde{f}*P_y)(x) \to \tilde{f}(x) \quad \text{a.e.}$$

as $y \to 0^+$. We shall see that this is also true if $f \in L^p$, $1 < p < \infty$.

THEOREM 2: (M. Riesz) Let $f \in L^p(E^1)$, $1 < p < \infty$, and

$$F(z) = F(x+iy) = U(x,y)+iV(x,y)$$

be as above. Then, for any fixed $y > 0$,

a) $$\int_{-\infty}^{\infty} |U(x,y)|^p \, dx \leq \int_{-\infty}^{\infty} |f(x)|^p \, dx$$

b) $$\int_{-\infty}^{\infty} |V(x,y)|^p \, dx \leq A_p \int_{-\infty}^{\infty} |f(x)|^p \, dx$$

c) $$\int_{-\infty}^{\infty} |F(x+iy)|^p \, dx \leq A_p \int_{-\infty}^{\infty} |f(x)|^p \, dx$$

where A_p is a positive constant dependent only on p.

Proof: a) is an immediate consequence of Young's Theorem, since

$$\|U(x,y)\|_p = \|(f*P_y)(x)\|_p \leq \|f\|_p \|P_y\|_1 = \|f\|_p.$$

In fact a) holds for all p such that $1 \leq p \leq \infty$.

Since $F = U + iV$, then c) follows from a) and b) by Minkowski's Inequality. So it remains to prove b). The following proof of b) is due to Calderon.

It suffices to prove the inequality for functions f with finite support. In fact for a general $f \in L^p$ we form the truncations, $f_n(x) = f(x)$ if $|x| \leq n$ and $f_n(x) = 0$ elsewhere, and let $V_n = f_n * \tilde{P}_y$. But $V_n(x,y) \to V(x,y)$ as $n \to \infty$, because if $q = p/(p-1)$ then by Hölder's Inequality

$$|V_n - V| \leq \|f_n - f\|_p \|\tilde{P}_y\|_q = C\|f_n - f\|_p \to 0$$

since $f_n \to f$ in L^p. So if we have that

$$\int_{-\infty}^{\infty} |V_n(x,y)|^p \, dx \leq A_p \int_{-\infty}^{\infty} |f_n(x)|^p \, dx \leq A_p \int_{-\infty}^{\infty} |f(x)|^p \, dx$$

then it follows by Fatou's Lemma that

$$\int_{-\infty}^{\infty} |V(x,y)|^p \, dx \leq A_p \int_{-\infty}^{\infty} |f(x)|^p \, dx.$$

Next, we may assume that f is real valued, by considering separately its real and imaginary parts. Moreover we may assume that $f(x) \geq 0$, by splitting f into its positive and negative parts: $f = f^+ - f^-$.

So we suppose that f has finite support and that $f(x) \geq 0$.

Since $P_y(x) > 0$ then, excluding the trivial case when f is identically zero, we have that $U(x,y) = (f*P_y)(x) > 0$.

Let $w = u + iv$ be any complex number with $u > 0$; then if $1 < p < 3$ there exist positive constants c_1 and c_2, dependent on p, such that

(1) $$|v|^p \leq c_1 u^p - c_2 \mathrm{Re}(w^p).$$

In fact, by homogeneity, we may assume that $w = e^{i\theta}$, where $0 \leq |\theta| < \frac{\pi}{2}$ since $u = \cos\theta > 0$. Then $w^p = \cos p\theta + i \sin p\theta$, so we must show that

(1') $$|\sin\theta|^p \leq c_1 (\cos\theta)^p - c_2 \cos p\theta.$$

Clearly $|\sin\theta|^p < 1$ if $|\theta| < \frac{\pi}{2}$. But for each p, with $1 < p < 3$, there exist a sufficiently small $\varepsilon > 0$ and a sufficiently large $c_2 > 0$ such that $-c_2 \cos p\theta \geq 1$ if $\frac{\pi}{2} - \varepsilon < |\theta| < \frac{\pi}{2}$. On the other hand if $|\theta| \leq \frac{\pi}{2} - \varepsilon$, we can now choose c_1 so large that $c_1(\cos\theta)^p \geq c_2 + 1$. Then the right side of (1') is again ≥ 1 and hence the inequality is proved.

Applying now inequality (1) to $w = F(z) = U + iV$ and integrating with respect to x we obtain

(2) $$\int_{-\infty}^{\infty} |V(x,y)|^p dx \leq c_1 \int_{-\infty}^{\infty} U(x,y)^p dx - c_2 \int_{-\infty}^{\infty} \mathrm{Re}\{F(x+iy)^p\} dx.$$

However, the last term in (2) vanishes. In fact $F(z)^p$ is holomorphic in $y = \mathrm{Im}(z) > 0$, and since

$$F(z) = \frac{1}{\pi i} \int_{-a}^{a} \frac{f(t)}{t-z}\, dt = \mathcal{O}\left(\frac{1}{|z|}\right)$$

we have that $F(z)^p = \mathcal{O}(|z|^{-p})$ where $p > 1$. Hence, by Cauchy's Theorem, if we integrate over a semicircular contour Γ at height y and let the radius of the contour tend to infinity we see that

$$\int_{-\infty}^{\infty} F(x+iy)^p\, dx = 0.$$

So, in particular, the real part of this integral must vanish.

Therefore, letting $C_1 = A_p$ and using estimate a) we obtain from (2) that

$$\int_{-\infty}^{\infty} |V(x,y)|^p\, dx \leq A_p \int_{-\infty}^{\infty} U(x,y)^p\, dx \leq A_p \int_{-\infty}^{\infty} f(x)^p\, dx.$$

Hence we have proved estimate b) for $1 < p < 3$, so, in particular for every p such that $1 < p \leq 2$. But the validity of the estimate for the estimate for the remaining values of $p < \infty$ is a special case of the following duality argument.

CLAIM: If b) holds for some p, $1 < p < \infty$, then it also holds for the conjugate exponent $q = p/(p-1)$ and with the same constant $A_q = A_p$.

Let $y > 0$ be fixed, $f \in L^q$ and $V(x,y) = (f*\tilde{P}_y)(x)$. Then, by the Converse of Hölder,

$$\left(\int_{-\infty}^{\infty} |V(x,y)|^q\, dx \right)^{1/q} = \sup_{\|g\|_p = 1} \left| \int_{-\infty}^{\infty} V(x,y) g(x)\, dx \right|$$

where we may assume that g, like f, has finite support. Interchanging the order of integration and recalling that $\tilde{P}_y$ is odd we have that

$$\int_{-\infty}^{\infty} V(x,y)g(x)\,dx = \int_{-\infty}^{\infty} g(x)\left\{\int_{-\infty}^{\infty} f(t)\tilde{P}_y(x-t)\,dt\right\}dx =$$

$$= \int_{-\infty}^{\infty} f(t)\left\{-\int_{-\infty}^{\infty} g(x)\tilde{P}_y(t-x)\,dx\right\}dt =$$

$$= -\int_{-\infty}^{\infty} f(t)W(t,y)\,dt$$

where $W(t,y)$ is the Conjugate Poisson Integral of g, Hence, using Hölder's Inequality and then estimate b) we obtain

$$\|V(x,y)\|_q = \sup_{\|g\|_p=1} \left|\int_{-\infty}^{\infty} V(x,y)g(x)\,dx\right| =$$

$$= \sup_{\|g\|_p=1} \left|\int_{-\infty}^{\infty} f(t)W(t,y)\,dt\right| \leqslant$$

$$\leqslant \|f\|_q \left(\int_{-\infty}^{\infty} |W(t,y)|^p\,dt\right)^{1/p} \leqslant$$

$$\leqslant \|f\|_q\, A_p\, \|g\|_p = A_p \|f\|_q$$

So the Claim is proved and the proof of the theorem is complete.

Q.E.D.

THEOREM 3: (M. Riesz) Let $f \in L^p(E^1)$, $1 < p < \infty$, and consider the truncated Hilbert transform

$$\tilde{f}_\varepsilon(x) = \frac{1}{\pi}\int_{|t|>\varepsilon} \frac{f(x-t)}{t}\,dt = \frac{1}{\pi}\int_{|x-t|>\varepsilon} \frac{f(t)}{x-t}\,dt.$$

Then,

a) $\|\tilde{f}_\varepsilon\|_p \leqslant A_p\|f\|_p$.

b) There exists a function $\tilde{f} \in L^p$ such that $\|\tilde{f}_\varepsilon - f\|_p \to 0$ as $\varepsilon \to 0$.

c) $\|\tilde{f}\|_p \leqslant A_p \|f\|_p$.

Proof: We recall from the proof of Th. 1 that

$$V(x,\varepsilon) - \tilde{f}_\varepsilon(x) = (f*K_\varepsilon)(x)$$

where $K_\varepsilon(t) = \frac{1}{\varepsilon} K(\frac{t}{\varepsilon})$ is an integrable kernel. Hence, by Young's Theorem,

(1) $$\|V(x,\varepsilon) - \tilde{f}_\varepsilon(x)\|_p \leq C\|f\|_p$$

where $C = \|K\|_1$.

On the other hand, from estimate b) of Th. 2 we have that for every $\varepsilon > 0$

(2) $$\|V(x,\varepsilon)\|_p \leq A_p\|f\|_p.$$

Therefore, by Minkowski's Inequality,

$$\|\tilde{f}_\varepsilon\|_p \leq \|\tilde{f}_\varepsilon(x) - V(x,\varepsilon)\|_p + \|V(x,\varepsilon)\|_p \leq (C+A_p)\|f\|_p$$

so a) is proved.

To prove b) we must show that, as $\varepsilon \to 0$, $\{\tilde{f}_\varepsilon\}$ is a Cauchy sequence in L^p. First, we verify that this is the case for functions belonging to a dense subset of L^p.

Let $g, h \in C_o^1(E^1)$ be continuously differentiable functions with compact support and suppose that $h(x)$ is even and $h(x) = 1$ near the origin. Now,

(i) $$\tilde{g}_\varepsilon(x) = \frac{1}{\pi} \int_{|x-t|>\varepsilon} \frac{g(t)}{x-t}\, dt.$$

Moreover, since $\int_{|x-t|>\varepsilon} \frac{h(x-t)}{x-t}\, dt = 0$, we have that

(ii) $$\tilde{g}_\varepsilon(x) = \frac{1}{\pi} \int_{|x-t|>\varepsilon} \frac{g(t)-g(x)h(x-t)}{x-t}\, dt.$$

Let $(-a,a)$ contain the support of g. Then if $|x| \geq 2a$,

$\big|\, |x|-|t|\, \big| \geq |x| - \frac{1}{2}|x| = \frac{1}{2}|x|$ for all $t \in (-a,a)$, so by (i)

$$|\tilde{g}_\varepsilon(x)| \leq \int_{|x-t|>\varepsilon} \frac{|g(t)|}{|x-t|}\, dt \leq \int_{-a}^{a} \frac{|g(t)|}{|\,|x|-|t|\,|}\, dt \leq \frac{C}{|x|} .$$

On the other hand, if $|x| < 2a$, $|g(t) - g(x)h(x-t)| < M|x-t|$, and since g and h have finite support, we have from (ii) that

$$|\tilde{g}_\varepsilon(x)| \leq B.$$

Hence, combining the two estimates, we have that for all x

(3) $$|\tilde{g}_\varepsilon(x)| \leq \frac{A}{1+|x|}$$

where the last function belongs to L^p for every $p > 1$.

But as $\varepsilon \to 0$, $\tilde{g}_\varepsilon(x)$ converges pointwise to

$$\tilde{g}(x) = \frac{1}{\pi} \int_{-\infty}^{\infty} \frac{g(t)-g(x)h(x-t)}{x-t}\, dt$$

so by (3) and Lebesgue's Theorem of Dominated Convergence it follows that $\tilde{g}_\varepsilon \to \tilde{g}$ in L^p norm; in particular, $\{\tilde{g}_\varepsilon\}$ is a Cauchy sequence in L^p.

Since C_o^1 is dense in L^p, then given any $f \in L^p$ and $\delta > 0$ there is a function $g \in C_o^1$ such that $\|f-g\|_p < \delta$. Now,

$$\tilde{f}_\varepsilon - \tilde{f}_\eta = (\tilde{f}_\varepsilon - \tilde{g}_\varepsilon) + (\tilde{g}_\varepsilon - \tilde{g}_\eta) + (\tilde{g}_\eta - \tilde{f}_\eta)$$

so by Minkowski's Inequality and estimate a), it follows that

$$\|\tilde{f}_\varepsilon - \tilde{f}_\eta\|_p \leq \|\widetilde{(f-g)}_\varepsilon\|_p + \|\tilde{g}_\varepsilon - \tilde{g}_\eta\|_p + \|\widetilde{(g-f)}_\eta\|_p \leq$$

$$\leq 2A_p\delta + \|\tilde{g}_\varepsilon - \tilde{g}_\eta\|_p$$

and since $\{\tilde{g}_\varepsilon\}$ is a Cauchy sequence in L^p, then so is $\{\tilde{f}_\varepsilon\}$. Therefore, there exists a function $\tilde{f}$ in L^p such that $\tilde{f}_\varepsilon \to \tilde{f}$ in L^p norm as $\varepsilon \to 0$.

Finally, a) and b) imply c), by continuity of the norm.

Q.E.D.

THEOREM 4: If $f \in L^p(E^1)$, $1 < p < \infty$, then

$$\tilde{f}(x) = \lim_{\varepsilon \to 0} \tilde{f}_\varepsilon(x) = \lim_{\varepsilon \to 0} \frac{1}{\pi} \int_{|x-t|>\varepsilon} \frac{f(t)}{x-t}\, dt$$

exists almost everywhere.

Proof: By the preceding theorem $\tilde{f} \in L^p$ so, by Young's Theorem, its Poisson Integral $(\tilde{f}*P_\varepsilon)(x)$ exists a.e. and belongs to L^p. By Th. 2, the Conjugate Poisson integral of f, $V(x,\varepsilon) = (f*\tilde{P}_\varepsilon)(x)$ also belongs to L^p. We claim that these two functions coincide, i.e.

$$(*) \qquad \int_{-\infty}^{\infty} \tilde{f}(t) P_\varepsilon(x-t)\, dt = \int_{-\infty}^{\infty} f(t) \tilde{P}_\varepsilon(x-t)\, dt.$$

We saw earlier that (*) is valid for all $f \in L^2$, so in particular is valid on C_0^1. For $f \in L^p$, choose a sequence $f_n \in C_0^1$ which converges to f in L^p norm, as $n \to \infty$. Then, by Th. 3. $\tilde{f}_n \to \tilde{f}$ in L^p norm, since the Hilbert transform is continuous in L^p, $1 < p < \infty$.

Now (*) is valid on the f_n, and as $n \to \infty$ we have, by Hölder's inequality with $q = p/(p-1)$, that

$$\left| \int_{-\infty}^{\infty} [f_n(t) - f(t)] \tilde{P}_\varepsilon(x-t)\, dt \right| \leq \|f_n - f\|_p \, \|\tilde{P}_\varepsilon\|_q \to 0$$

and similarly for the left side of (*). Therefore (*) holds for every $f \in L^p$, $1 < p < \infty$.

From Th. 7 of Chapter I, $(\tilde{f}*P_\varepsilon)(x) \to \tilde{f}(x)$ a.e., as $\varepsilon \to 0$, so by (*)

$$V(x,\varepsilon) = (f*\tilde{P}_\varepsilon)(x) = (\tilde{f}*P_\varepsilon)(x) \to \tilde{f}(x) \text{ a.e. .}$$

But we recall from the proof of Th. 1 that, even with $f \in L^p$,

$$V(x,\varepsilon) - \tilde{f}_\varepsilon(x) = (f*K_\varepsilon)(x) \to f(x) \int_{-\infty}^{\infty} K(t)\, dt = 0$$

for almost every x, i.e. $V(x,\varepsilon)$ and $\tilde{f}_\varepsilon(x)$ are equiconvergent. Hence we must have that $\tilde{f}_\varepsilon(x) \to \tilde{f}(x)$ a.e., as $\varepsilon \to 0$.

Q.E.D.

COROLLARY 2: (Inversion Formula) If $f \in L^p(E^1)$, $1 < p < \infty$, then

$$\tilde{\tilde{f}} = - f.$$

Proof: We saw that this formula is valid in L^2 hence, in particular, on C_o^1 which is dense in L^p. Since both sides are continuous operations in L^p, as we proved for the Hilbert transform and as is obvious for scalar multiplication, the formula holds by passage to the limit.

Q.E.D.

COROLLARY 3: If $f \in L^p(E^1)$, $1 < p < \infty$, and $g \in L^q(E^1)$, $q = p/(p-1)$, then

(i) $$\int_{-\infty}^{\infty} \tilde{f}\, \tilde{g}\, dx = \int_{-\infty}^{\infty} f\, g\, dx$$

(ii) $$\int_{-\infty}^{\infty} \tilde{f}\, g\, dx = -\int_{-\infty}^{\infty} f\, \tilde{g}\, dx$$

Proof: The above integrals exist, by Hölder's inequality, since the Hilbert transform preserves the L^p spaces. Moreover (ii) follows readily from (i) and the preceding corollary.

Since (i) is valid on C_o^1, we take sequences $\{f_n\}$ and $\{g_n\}$ in C_o^1 such that $f_n \to f$ in L^p norm and $g_n \to g$ in L^q norm. Then $\tilde{f}_n \to \tilde{f}$ in L^p and $\tilde{g}_n \to \tilde{g}$ in L^q; moreover

$$\int \tilde{f}_n\, \tilde{g}_n\, dx = \int f_n\, g_n\, dx.$$

But, by Hölder's inequality, these integrals are continuous bilinear functionals on $L^p \times L^q$, so (i) follows by passage to the limit.

Q.E.D.

Summing up, we have proved that for $f \in L^p(E^1)$, $1 < p < \infty$, the Hilbert transform

$$(Hf)(x) = \tilde{f}(x) = \text{p.v.}\ \frac{1}{\pi} \int_{-\infty}^{\infty} \frac{f(t)}{x-t}\, dt$$

exists pointwise a.e. and also as a limit in L^p norm. It is a continuous operator in L^p, $1 < p < \infty$, which is invertible, in fact $H^{-1} = -H$, and skew-adjoint, i.e. $H^* = -H$.

We conclude this chapter with some results about the Hilbert transform in the space $L^1(E^1)$.

If f is an integrable function we denote by $e(y)$, $y > 0$, its distribution function:

$$e(y) = |\{x: |f(x)| > y\}|.$$

Then we have Tchebishev's Inequality

$$e(y) \leq \frac{1}{y} \int |f(x)|\, dx = \frac{\|f\|_1}{y}$$

whoose proof (geometrically evident) is left as an exercise for the reader. Using this inequality one obtains, considering Lebesgue sums and then integrating by parts, the familiar formula

$$\int |f(x)|\, dx = -\int_0^{\infty} y\, de(y) = \int_0^{\infty} e(y)\, dy.$$

More generally, for any $p > 0$,

$$\int |f(x)|^p\, dx = -\int_0^{\infty} y^p\, de(y) = p \int_0^{\infty} y^{p-1}\, e(y)\, dy.$$

We note that the above formulas are valid for measurable functions on E^n, and that the distribution function $e(y)$ can be defined for any measurable function f, integrable or not.

REMARK: A linear operator $T: L^p \to L^p$, $1 \leq p < \infty$, is said to be of

<u>type</u> (p,p) if it is continuous, i.e.

$$\|Tf\|_p \leq A\|f\|_p.$$

A linear operator T, defined on L^p, is said to be of <u>weak type</u> (p,p) if it satisfies an estimate of the form

$$E(y) \leq A\left(\frac{\|f\|_p}{y}\right)^p, \qquad y > 0$$

where,

$$E(y) = |\{x: |Tf(x)| > y\}|$$

is the distribution function of Tf. By Tchebishev's inequality in L^p it is clear that an operator of type (p,p) is also of weak type (p,p).

We saw by examples that the Hilbert transform does not map L^1 into L^1, so a fortiori it is not of type (1,1). The following theorem, however, shows that $Hf = \tilde{f}$ is of weak type (1,1).

THEOREM 5: Let $f \in L^1(E^1)$; then $\tilde{f}(x)$ exists almost everywhere. If $E(y) = |\{x: |\tilde{f}(x)| > y > 0\}|$ is the distribution function of $\tilde{f}$, then

$$E(y) \leq A\,\frac{\|f\|_1}{y}$$

where $A > 0$ is a constant independent of y and f.

More generally,

THEOREM 6: Let F be a function of bounded variation on E^1 and V be its total variation. Then

$$g(x) = \text{p.v. } \frac{1}{\pi}\int_{-\infty}^{\infty} \frac{dF(t)}{x-t}\,dt$$

exists almost everywhere. Moreover, if $E(y) = |\{x: |g(x)| > y > 0\}|$ is the distribution function of g,

$$E(y) \leq A\,\frac{V}{y}\,.$$

We note that the function $\pi g_{\varepsilon}(x) = \int_{|x-t|>\varepsilon} \frac{dF(t)}{x-t}$ is defined for every x. In fact if K is any compact set in the complement of $[x-\varepsilon, x+\varepsilon]$ then the function $(x-t)^{-1}$ is continuous on K and so the Lebesgue-Stieltjes integral $\int_K \frac{dF(t)}{x-t}$ exists. To verify the convergence at infinity we observe that if, for example, $x + \varepsilon < a < b$ then

$$\left| \int_a^b \frac{dF(t)}{x-t} \right| \leq \int_a^b \frac{dV}{|x-t|} \leq \frac{V}{|x-a|} .$$

Next we observe that Th. 6 implies Th. 5. In fact if $f \in L^1(E^1)$ and

$$F(t) = \int_{-\infty}^{t} f(u)\, du$$

then $dF(t) = f(t)\,dt$ a.e., $F(t)$ is of bounded variation on E^1 and its total variation is

$$V = \int_{-\infty}^{\infty} |f(u)|\, du = \|f\|_1 .$$

For the proof of Th. 6, see Loomis, Bull. Amer. Math. Soc., 52, (1946), pp. 1082-1086.

From Th. 5, we obtain the following

COROLLARY 4: If $f_n \in L^1(E^1)$ and $f_n \to f$ in L^1 norm, then $\tilde{f}_n \to \tilde{f}$ in measure. For some subsequence $\{f_{n_k}\}$, $\tilde{f}_{n_k} \to \tilde{f}$ almost everywhere.

Proof: Applying Th. 5 to $f_n - f$ we have that for every $\varepsilon > 0$

$$|\{x: |\tilde{f}_n(x) - \tilde{f}(x)| > \varepsilon\}| \leq \frac{A}{\varepsilon} \|f_n - f\|_1 \to 0$$

as $n \to \infty$, hence $\tilde{f}_n \to \tilde{f}$ in measure.

Moreover, from any sequence of functions which converges in measure one can select a subsequence converging almost everywhere.

Q.E.D.

Finally we shall prove that if $f \in L^1$ then $\tilde{f}$ belongs <u>locally</u> to L^p for every p such that $0 < p < 1$.

THEOREM 7: (Kolmogorov) Let B be a bounded set in E^1 and $0 < \alpha < 1$. If $f \in L^1(E^1)$ then $\tilde{f} \in L^{1-\alpha}(B)$.

Proof: Let $E(y) = |\{x: |\tilde{f}(x)| > y > 0\}|$ be the distribution function of $\tilde{f}$, and set $\|f\|_1 = V$. Then, by Th. 5,

$$E(y) \leq A \frac{V}{y} .$$

If $e(y) = |\{x: x \in B, |\tilde{f}(x)| > y > 0\}|$ is the distribution function of the restriction of $\tilde{f}$ to B, then clearly

(i) $e(y) \leq |B|$

(ii) $e(y) \leq E(y) \leq A \frac{V}{y}$.

We must prove that the integral $I = \int_B |\tilde{f}(x)|^{1-\alpha} dx$ is finite.

Introducing e(y) and integrating by parts we obtain

$$I = -\int_0^\infty y^{1-\alpha} de(y) = [-y^{1-\alpha} e(y)]_0^\infty + \int_0^\infty (1-\alpha) y^{-\alpha} e(y)\, dy =$$

$$= (1-\alpha)\int_0^\infty y^{-\alpha} e(y)\, dy$$

since, using (i) and (ii),

as $y \to \infty$ $\quad |y^{1-\alpha} e(y)| \leq AVy^{-\alpha} \to 0$, and

as $y \to 0$ $\quad |y^{1-\alpha} e(y)| \leq |B| y^{1-\alpha} \to 0$.

By estimates (i) and (ii) again, we have that for any y_o, $0 < y_o < \infty$,

$$I \leq (1-\alpha)|B| \int_0^{y_o} y^{-\alpha} dy + (1-\alpha) AV \int_{y_o}^\infty y^{-1-\alpha} dy =$$

$$= |B| y_o^{1-\alpha} + \frac{1-\alpha}{\alpha} AVy_o^{-\alpha} \equiv g(y_o)$$

which shows that I is finite and proves the theorem. However, we can also obtain a sharp estimate of I if we choose y_o so that $g(y_o)$ is a minimum. Since

$$g'(y) = (1-\alpha)|B|y^{-\alpha} - (1-\alpha)AVy^{-1-\alpha} =$$

$$= (1-\alpha)y^{-1-\alpha}(y|B| - AV)$$

we choose $y_o = \frac{AV}{|B|}$, and our minimum value is

$$g(y_o) = g\left(\frac{AV}{|B|}\right) = \frac{(AV)^{1-\alpha}}{\alpha}|B|^{\alpha}.$$

Hence, recalling the definition of I and of V, we obtain

$$\int_B |\tilde{f}(x)|^{1-\alpha} dx \leqslant \frac{A^{1-\alpha}}{\alpha} |B|^{\alpha} \|f\|_1^{1-\alpha} .$$

Q.E.D.

EXERCISES

1. Let $h \in L^{\infty}(E^n)$ and $H: L^2(E^n) \to L^2(E^n)$ be the bounded linear operator given by $(Hf)(x) = h(x)f(x)$. Prove that $\|H\| = \|h\|_{\infty}$. (Hint: with $M = \|h\|_{\infty}$ show that, for any $\varepsilon > 0$, $\|H\| \geqslant M - \varepsilon$).
2. Define the Fourier transform $Ff = \hat{f}$ by the formula

$$\hat{f}(x) = \int f(y)\, e^{-2\pi i(x\cdot y)}\, dy$$

so that $F^* = F^{-1}$, i.e. F is a unitary transformation of $L^2(E^n)$. Prove that the class of operators $H: L^2(E^n) \to L^2(E^n)$ of the form

$$H = F^{-1} h(x) F \quad , \text{ where } h \in L^{\infty}(E^n),$$

is a commutative B*-algebra isomorphic and isometric to $L^{\infty}(E^n)$. (Hint: First show that these operators from a commutative algebra with adjoint, i.e. * algebra. Then show that the symbol map $\sigma: H \to h$ gives the desired isomorphism and that, using the previous exercise, σ is an isometry Finally, show that the operators form a Banach space, i.e. that the class is complete).

3. Let $H = F^{-1}hF$, $h \in L^{\infty}(E^n)$, $H: L^2(E^n) \to L^2(E^n)$, as in the previous exercise. Let iff = if and only if. Prove that

H is self-adjoint iff $h(x)$ is real valued,

H is skew-adjoint iff $h(x)$ is purely imaginary,

H is invertible iff $|h(x)| \geqslant \varepsilon > 0$,

H is unitary iff $|h(x)| = 1$,

H is positive definite iff $h(x) \geqslant \varepsilon > 0$,

where the various conditions on $h(x)$ are assumed to hold a.e.. Prove that the spectrum of H is essentially the closure of the range of $h(x)$.

4. Prove that the spectrum of the Hilbert transform in L^2 consists of the points $\{-i,i\}$ which are eigenvalues of infinite multiplicity.

5. Prove the analogue of Tchebishev's Inequality for functions $f \in L^p(E^n)$.

6. Let $f \in L^1(E^n)$, $f(x) \geqslant 0$, and $e(y) = |\{x: f(x) > y > 0\}|$.

Verify the following properties of $e(y)$:

a) $e(y)$ is monotone decreasing (i.e. non-increasing) and continuous from the right.

b) $|\{x: \alpha < f(x) \leqslant \beta\}| = e(\alpha) - e(\beta)$.

c) $|\{x: \alpha < f(x) < \beta\}| = e(\alpha) - e(\beta-0)$

d) $|\{x: \alpha \leqslant f(x) \leqslant \beta\}| = e(\alpha-0) - e(\beta)$.

e) $e(y)$ is continuous at $y = \alpha$ if and only if $|\{x: f(x) = \alpha\}| = 0$.

f) $e(y)$ is bounded if and only if $f(x)$ has bounded support and $e(y)$ has bounded support if and only if $f(x)$ is bounded.

g) $\displaystyle\int_{E^n} f(x)\,dx = -\int_0^{\infty} y\,de(y) = \int_0^{\infty} e(y)\,dy.$

h) $\displaystyle\int_{E^n} f(x)^p\,dx = -\int_0^{\infty} y^p de(y) = p\int_0^{\infty} y^{p-1}\,e(y)\,dy$, for any $p > 0$.

CHAPTER IV

SINGULAR INTEGRALS IN E^n

§1. EXISTENCE. ODD KERNELS

In this chapter we consider the extension to E^n of the results in Chapter III about the Hilbert transform. For this reason, the singular integrals with odd kernels are sometimes called Hilbert transforms in E^n. The basic work in this theory was done by Mihlin and by Calderon and Zygmund. We shall follow here a method developed by Calderon and Zygmund.

A kernel $K(x)$ is said to be (positively) homogeneous of degree α if, for all $\lambda > 0$ and $x \in E^n$, $K(\lambda x) = \lambda^{\alpha} K(x)$.

If $x \neq 0$, we denote by x' its projection on the unit sphere $\Sigma = \{x: |x| = 1\}$, that is $x' = \frac{x}{|x|}$. So, if $K(x)$ is homogeneous of degree α, we have that

$$K(x) = |x|^{\alpha} K(\frac{x}{|x|}) = |x|^{\alpha} K(x').$$

The function $\Omega(x) = K(\frac{x}{|x|})$ is homogeneous of degree zero, so $\Omega(x) = \Omega(x')$ can be viewed as a function on Σ. This function is sometimes called the characteristic of the kernel K.

Consider now a transformation of the form $f \to f*K$, where $K(x)$ is homogeneous of negative degree $-\alpha$ and $f*K$ is the convolution

$$(f*K)(x) = \int_{E_n} f(y) K(x-y)\, dy.$$

We have three distinct cases:

(i) $0 < \alpha < n$. The integral is said to be <u>weakly singular</u>.

(ii) $\alpha = n$. The integral is said to be <u>singular</u>.

(iii) $\alpha > n$. The integral is said to be <u>hypersingular</u>.

Weakly singular integrals have been studied extensively.

Examples of such integrals are the Newtonian Potentials $f*|x|^{-n+2}$, $n > 2$, and more generally the Riesz Potentials, $f*|x|^{-\alpha}$, $0 < \alpha < n$. It is known that they determine completely continuous (i.e. compact) operators on the spaces $L^p(E)$, where E is a bounded domain in E^n and $1 \leq p \leq \infty$.

Hypersingular integrals are relatively unknown. They have appeared however in connection with the study of smoothness properties of functions.

We shall confine our attention to the study of singular integrals. One example we have already seen is the Hilbert transform. Another important example arises if we try to calculate the second order derivatives of a Newtonian Potential by differentiating, formally, under the integral sign. For a general description of the subject of singular integrals and several of its applications we refer the reader to the lectures of Calderon, Bull. Amer. Math. Soc., 72 (1966) pp. 427-465, where an extensive bibliography can be found.

Consider a singular integral $\tilde{f}(x) = (f*K)(x)$, where $K(x)$ is homogeneous of degree $-n$ and the integral is taken in the principal value sense:

$$\tilde{f}(x) = \text{p.v.} \int f(y) K(x-y)\, dy = \lim_{\varepsilon \to 0+} \tilde{f}_\varepsilon(x)$$

$$\tilde{f}_\varepsilon(x) = \int_{|x-y|>\varepsilon} f(y) K(x-y)\, dy.$$

We note that $\tilde{f}_\varepsilon$ can also be written as a convolution: $\tilde{f}_\varepsilon = f*K_\varepsilon$, where $K_\varepsilon(x) = K(x)$ if $|x| > \varepsilon$ and $K_\varepsilon(x) = 0$ otherwise.

We shall assume from the start that $\Omega(x) = K(x')$ is integrable on Σ.

LEMMA 1: If $f \in L^p(E^n)$, $1 \leq p < \infty$, and $\Omega \in L^1(\Sigma)$, then $\tilde{f}_\varepsilon(x)$ exists almost everywhere.

Proof: Let

$$I(x) = \int_{|x-y|>\varepsilon} |f(y)|\ |K(x-y)|\ dy = \int_{|y|>\varepsilon} |K(y)|\ \ |f(x-y)|\,dy.$$

Since $|\tilde{f}_\varepsilon(x)| \leq I(x)$, if we show that $I(x)$ is locally integrable this will imply that $I(x)$, and hence $\tilde{f}_\varepsilon(x)$, is finite a.e. .

So, if B is any bounded set in E^n, we will show that the integral $J = \int_B I(x)dx$ is finite.

Letting $r = |y|$, $y' = y/r$, dy' the surface element on Σ, we have that

$$J = \int_B dx \int_{|y|>\varepsilon} \frac{|\Omega(y')|}{|y|^n} |f(x-y)|dy =$$

$$= \int_B dx \int_\Sigma |\Omega(y')| \left[\int_\varepsilon^\infty |f(x-ry')|r^{-1}dr\right] dy' =$$

$$= \int_\Sigma |\Omega(y')| \left\{\int_B dx \left[\int_\varepsilon^\infty |f(x-ry'|r^{-1}dr\right]\right\} dy'.$$

If $p = 1$, then

$$\int_B dx \left[\int_\varepsilon^\infty |f(x-ry')|r^{-1}dr\right] \leq \frac{1}{\varepsilon} \int_B dx \int_{-\infty}^\infty |f(x-ry')|dr \leq \frac{C}{\varepsilon} \|f\|_1$$

where we have integrated first along the line $x - ry'$ and then over B, and where C denotes the diameter of B. Therefore

$$J \leq \frac{C}{\varepsilon} \|\Omega\|_1 \|f\|_1 \quad \text{, where } \|\Omega\|_1 = \int_\Sigma |\Omega(y')|dy'.$$

If $1 < p < \infty$, and $q = p/(p-1)$ is the conjugate exponent then, by Hölder's inequality,

$$\int_\varepsilon^\infty |f(x-ry')|r^{-1}\,dr \leq A_{\varepsilon,q}\left(\int_\varepsilon^\infty |f(x-ry')|^p dr\right)^{1/p}$$

where $A_{\varepsilon,q}$ is a constant dependent on ε and q. Using this extimate and applying again Hölder's inequality we obtain

$$\int_B dx \int_\varepsilon^\infty |f(x-ry')| r^{-1} dr \leq A_{\varepsilon,q} \int_B dx \left(\int_\varepsilon^\infty |f(x-ry')|^p dr \right)^{1/p} \leq$$

$$\leq A_{\varepsilon,q} |B|^{1/q} \left(\int_B dx \int_\varepsilon^\infty |f(x-ry')|^p dr \right)^{1/p} \leq$$

$$\leq A'_{\varepsilon,q} |B|^{1/q} \|f\|_p .$$

Therefore

$$J \leq A'_{\varepsilon,q} |B|^{1/q} \|\Omega\|_1 \|f\|_p .$$

Q.E.D.

LEMMA 2: If $f \in L^p(E^n)$, $1 \leq p \leq \infty$, and Ω is bounded, then $\tilde{f}_\varepsilon(x)$ exists everywhere.

Proof: If $|\Omega(y')| \leq M$, then

$$|\tilde{f}_\varepsilon(x)| \leq \int_{|y|>\varepsilon} |f(x-y)| \frac{|\Omega(y')|}{|y|^n} dy \leq M \int_{|y|>\varepsilon} |f(x-y)| \, |y|^{-n} dy$$

and if q is the conjugate exponent of p, then, by Hölder's inequality, the last integral is majorized by

$$\|f\|_p \left(\int_{|y|>\varepsilon} \frac{dy}{|y|^{nq}} \right)^{1/q} < \infty$$

with the obvious modifications if $p = 1$ and $q = \infty$.

Q.E.D.

In order to obtain the existence of $\tilde{f}(x) = \lim_{\varepsilon \to 0} \tilde{f}_\varepsilon(x)$, at least for functions belonging to a dense subset of L^p, we assume:

(i) $\Omega \in L^1(\Sigma)$

(ii) $\int_{\Sigma} \Omega(x') \, dx' = 0.$

Condition (i), as we saw, is natural. The necessity of condition (ii) will be seen in a remark below. This second assumption is usually expressed by saying that Ω has mean value zero on Σ.

We write $f \in \Lambda_\alpha$ if f satisfies a Hölder condition of order α.

THEOREM 1: Let $f \in L^p \cap \Lambda_\alpha$, $1 \leq p < \infty$ and $0 < \alpha \leq 1$, and suppose that Ω satisfies the condition (i) and (ii) above. Then $\tilde{f}(x) = \lim_{\varepsilon \to 0} \tilde{f}_\varepsilon(x)$ exists for almost every x. If, in addition, Ω is bounded then $\tilde{f}(x)$ exists everywhere.

Proof: since $f \in \Lambda_\alpha$, there are positive constants δ and C such that

$$|f(x+y)-f(x)| \leq C|y|^\alpha \quad , \text{ if } |y| \leq \delta .$$

By Lemma 1, we know that, for all $\varepsilon > 0$, $\tilde{f}_\varepsilon(x)$ exists a.e. . In particular, if $0 < \varepsilon < \delta$, we can write, for all x where $\tilde{f}_\delta$ is defined (i.e. almost everywhere, and everywhere if Ω is bounded)

$$\tilde{f}_\varepsilon(x) = \int_{\varepsilon \leq |y| \leq \delta} f(x-y) \frac{\Omega(y)}{|y|^n} dy + \tilde{f}_\delta(x) .$$

Since Ω has mean value zero on Σ, we have that

$$\int_{\varepsilon \leq |y| \leq \delta} \frac{\Omega(y)}{|y|^n} dy = \int_\varepsilon^\delta \frac{dr}{r} \int_\Sigma \Omega(y') \, dy' = 0$$

and therefore we can write

$$\tilde{f}_\varepsilon(x) = \int_{\varepsilon \leq |y| \leq \delta} [f(x-y)-f(x)] \frac{\Omega(y)}{|y|^n} dy + \tilde{f}_\delta(x) .$$

Now we can let $\varepsilon \to 0$ and the integral is convergent since the integrand $g(y) = [f(x-y)-f(x)]\Omega(y)|y|^{-n}$ is absolutely integrable on the ball $|y| \leq \delta$.

In fact

$$|g(y)| \leq C|\Omega(y)|\ |y|^{\alpha-n}, \qquad \text{if } |y| \leq \delta,$$

and

$$\int_{|y|\leq\delta} \frac{|\Omega(y)|}{|y|^{n-\alpha}}\, dy = \left(\int_0^\delta \frac{dr}{r^{1-\alpha}}\right)\left(\int_\Sigma |\Omega(y')|\,dy'\right) < \infty$$

since Ω is integrable on Σ.

Hence $\tilde{f}(x) = \lim_{\varepsilon\to 0} \tilde{f}_\varepsilon(x)$ exists a.e., and everywhere if Ω is bounded.

Q.E.D.

REMARK: (Necessity of the mean value zero condition).

Let $f \in C_o^1(E^n)$, $f(x) = 1$ on a closed ball B, and suppose that Ω is bounded. Then $\tilde{f}(x)$ exists at any point in the interior of B if and only if Ω has mean value zero on Σ.

In fact if x_o is in the interior of B we can choose $\delta > 0$ so that $x_o - y \in B$ for all $|y| \leq \delta$. Then if $0 < \varepsilon < \delta$ we have, as in the proof of Th. 1, that

$$\tilde{f}_\varepsilon(x_o) = \int_{\varepsilon\leq|y|\leq\delta} \frac{\Omega(y)}{|y|^n}\, dy + \tilde{f}_\delta(x_o)$$

where $\tilde{f}_\delta(x_o)$ exists by Lemma 2. However the integral

$$\int_{\varepsilon\leq|y|\leq\delta} \frac{\Omega(y)}{|y|^n}\, dy = \int_\varepsilon^\delta \frac{dr}{r} \int_\Sigma \Omega(y')\, dy'$$

will not converge as $\varepsilon \to 0$ unless of course $\int_\Sigma \Omega(y')\, dy' = 0$.

EXAMPLES:

1. Case $n = 1$.

Here the unit sphere consists of the points $\{-1,1\}$ so, in view of condition (ii), the mean value zero condition, we must have

$$K(x) = \frac{\Omega(x')}{|x|} = \frac{C \operatorname{sgn} x}{|x|} = \frac{C}{x}$$

where C is complex number. In other words, our singular integrals in E^1 are scalar multiples of the Hilbert transform.

2. Case n = 2.

Denoting by $z = re^{i\theta}$ the points of E^2, in polar form, we have

$$K(z) = \frac{\Omega(\theta)}{r^2}$$

where Ω is periodic of period 2π and integrable, condition (i). The Fourier series of Ω is

$$\Omega(\theta) \sim \sum_{\substack{-\infty \\ k\neq 0}}^{\infty} c_k e^{ik\theta} = \Sigma' c_k e^{ik\theta}$$

since $c_o = 0$ by condition (ii) So the Fourier series of K will be

$$K(z) \sim \frac{1}{r^2} \; \Sigma' \, c_k e^{ik\theta} \, .$$

In particular we have the kernels $K_k(z) = \frac{e^{ik\theta}}{r^2}$ where k is any non-zero integer. Taking k = -2 we obtain the kernel $K(z) = \frac{1}{z^2}$. Its corresponding singular integral

$$\tilde{f}(z) = \text{p.v.} \int_{E^2} \frac{f(\zeta)}{(\zeta - z)^2} \, d\zeta$$

is called the Beurling transform.

3. Case $n \geqslant 2$.

If $x = (x_1, \ldots, x_n) \in E^n$, we have, for example, the kernels of M. Riesz:

$$K_j(x) = \frac{x_j}{|x|^{n+1}}, \quad j = 1,2, \ldots, n.$$

The corresponding singular integrals (principal value convolutions)

$$\tilde{f} = R_j f = f * K_j$$

are called the Riesz transforms.

We shall consider now the continuity in the spaces $L^p(E^n)$, $1 < p < \infty$, of singular integral operators $\tilde{f} = f*K = \lim_{\varepsilon \to 0} \tilde{f}_\varepsilon$ where K is an odd function homogeneous of degree -n. In other words, $\Omega(-x') = - \Omega(x')$, which clearly implies that Ω has mean value zero on Σ.

THEOREM 2: Let $f \in L^p(E^n)$, $1 < p < \infty$, and suppose that Ω is odd and integrable on Σ. Then,

a) $\|\tilde{f}_\varepsilon\|_p \leq B_p \|f\|_p$.

b) There exists a function $\tilde{f} \in L^p$ such that $\|\tilde{f}_\varepsilon - \tilde{f}\|_p \to 0$, as $\varepsilon \to 0$.

c) $\|\tilde{f}\|_p \leq B_p \|f\|_p$ where $B_p = \frac{\pi}{2} A_p \|\Omega\|_1$ and A_p is the constant of the theorem of M. Riesz.

Proof: To prove a) we use the Method of Rotation which reduces the problem to the one-dimensional case. Since Ω is odd,

$$\tilde{f}_\varepsilon(x) = \int_{|y|>\varepsilon} f(x-y) \frac{\Omega(y')}{|y|^n} dy = - \int_{|y|>\varepsilon} f(x+y) \frac{\Omega(y')}{|y|^n} dy$$

and we recall from the proof of Lemma 1 that this integral is absolutely convergent for almost every x. Taking semi-sums and introducing polar coordinates, $r = |y|$ and $dy = r^{n-1} dr\, dy'$, we obtain

$$\tilde{f}_\varepsilon(x) = \frac{1}{2} \int_{|y|>\varepsilon} [f(x-y) - f(x+y)] \frac{\Omega(y')}{|y|^n} dy =$$

$$= \frac{1}{2} \int_\Sigma \Omega(y') \left\{ \int_\varepsilon^\infty \frac{f(x-ry') - f(x+ry')}{r} dr \right\} dy' =$$

$$= \frac{1}{2} \int_{\Sigma} \Omega(y')F_{\varepsilon}(x,y')\, dy'.$$

Hence, by Minkowski's integral inequality,

$$(*) \qquad \|\tilde{f}_{\varepsilon}\|_p \leq \frac{1}{2} \int_{\Sigma} |\Omega(y')| \; \|F_{\varepsilon}(x,y')\|_p \, dy'.$$

Fixing $y' \in \Sigma$, we examine the integral

$$I = \|F_{\varepsilon}(x,y')\|_p^p = \int_{E^n} \left| \int_{\varepsilon}^{\infty} \frac{f(x-ry')-f(x+ry')}{r}\, dr \right|^p dx.$$

Let H be the hyperplane orthogonal to the line $L = \{ty'\colon t \text{ real}\}$. Then any $x \in E^n$ can be written uniquely in the form $x = z + ty'$ where $z \in H$ is the point of intersection of the line passing through x which is parallel to L. Expressing the integral I in terms of the (z,t) co-ordinates we obtain

$$I = \int_{H} dz \int_{-\infty}^{\infty} dt \left| \int_{\varepsilon}^{\infty} \frac{f[z+(t-r)y']-f[z+(t+r)y']}{r}\, dr \right|^p.$$

If we introduce the function of a single variable $g(s) = f(z+sy')$, the innermost integral becomes

$$\int_{\varepsilon}^{\infty} \frac{g(t-r)-g(t+r)}{r}\, dr = \pi\, \tilde{g}_{\varepsilon}(t)$$

where $\tilde{g}_{\varepsilon}$ is the truncated Hilbert transform of g. Therefore, by Th. 3 of Chapter III, we have that

$$I = \pi^p \int_{H} \|\tilde{g}_{\varepsilon}\|_p^p \, dz \leq \pi^p A_p^p \int_{H} \|g\|_p^p \, dz =$$

$$= \pi^p A_p^p \int_{H} dz \int_{-\infty}^{\infty} |f(z+sy')|^p \, ds = \pi^p A_p^p \|f\|_p^p$$

whence

$$\|F_{\varepsilon}(x,y')\|_p = I^{1/p} \leq \pi A_p \|f\|_p.$$

Substituting this estimate in the expression (*) above, we obtain

$$\|\tilde{f}_\varepsilon\|_p \leq \frac{\pi}{2} A_p \|\Omega\|_1 \|f\|_p$$

which is the desired result.

Estimate c) follows directly from a) and b), by continuity of the norm.

So it remains to prove b).

The proof of b) is almost identical to the one-dimensional case, i.e. Th. 3 of last chapter. It suffices to verify that the sequence (actually net) $\{\tilde{g}_\varepsilon\}$ is Cauchy, as $\varepsilon \to 0$, for $g \in C_o^1(E^n)$. We take a radial function $h \in C_o^1$ such that $h(x) = 1$ near the origin. Then, since Ω has mean value zero on Σ, we show that

$$\tilde{g}_\varepsilon(x) = \int_{|y|>\varepsilon} g(x-y) \frac{\Omega(y)}{|y|^n} dy = \int_{|y|>\varepsilon} [g(x-y)-g(x)h(y)] \frac{\Omega(y)}{|y|^n} dy$$

is pointwise convergent as $\varepsilon \to 0$, and that for some constant $A > 0$, independent of x and ε,

$$|\tilde{g}_\varepsilon(x)| \leq \frac{A}{1+|x|^n}$$

where the last function is in $L^p(E^n)$ for all $p > 1$. By Lebesgue's Theorem of Dominated Convergence, it follows that $\{\tilde{g}_\varepsilon\}$ is Cauchy in L^p.

Q.E.D.

REMARKS:

1. In the general case, we may decompose the kernel K into its even and odd parts: $K(x) = \frac{1}{2}[K(x)+K(-x)] + \frac{1}{2}[K(x)-K(-x)]$. For the odd part, we use Th. 2 above. For the even part, we shall obtain in §3 the following result.

THEOREM: Let $f \in L^p(E^n)$, $1 < p < \infty$, and suppose that Ω is even, $\int_\Sigma \Omega(x')dx' = 0$, and $\Omega \in L \log^+ L(\Sigma)$. Then, for some constant $B_p > 0$ the conclusions of Th. 2 hold. The condition $\Omega \in L \log^+ L(\Sigma)$, which means that

$$\int_\Sigma |\Omega(x')| \log^+ |\Omega(x')| dx' < \infty$$

cannot be weakened. In particular, $\Omega \in L^1(\Sigma)$ is not enough, but the assumption that $\Omega \in L^r(\Sigma)$, $r > 1$, suffices.

2. In connection with several applications, particularly to partial differential operators with variable coefficients, it is convenient to introduce singular integral operators with "variable" kernels:

$$K(x,z) = \frac{\Omega(x,z')}{|z|^n}$$

where $\Omega(x,z')$ is a function on $E^n \times \Sigma$ which is bounded in x.

Letting

$$\Omega^*(z') = \sup_x |\Omega(x,z')|$$

we obtain the following extension of Th. 2:

THEOREM 2': Let $f \in L^p(E^n)$, $1 < p < \infty$, and suppose that $\Omega(x,z')$ is an odd function of z' and that $\Omega^*(z')$ is integrable on Σ. Then

$$\tilde{f}_\varepsilon(x) = \int_{|x-y|>\varepsilon} f(y)K(x,x-y)\, dy$$

exists almost everywhere, and

a) $\|\tilde{f}_\varepsilon\|_p \leq B_p \|f\|_p$.

b) There exists a function $\tilde{f} \in L^p$ such that $\|\tilde{f}_\varepsilon - \tilde{f}\|_p \to 0$, as $\varepsilon \to 0$.

c) $\|\tilde{f}\|_p \leq B_p \|f\|_p$, where $B_p = \frac{\pi}{2} A_p \|\Omega^*\|_1$

The proof is identical to that of Th. 2, except that one uses the majorization $|\Omega(x,y')| \leq |\Omega^*(y')|$.

§ 2. L^2-THEORY. THE RIESZ TRANSFORMS

As we saw in the study of the Hilbert transform, the L^2 theory of singular integral operators is rendered much simpler by using the Fourier transform. Here we shall take the Fourier transform in the form

$$\hat{f}(x) = \int f(y) e^{-2\pi i (x\cdot y)} dy$$

which has the advantage of avoiding the introduction of normalizing factors and which defines a unitary transformation of $L^2(E^n)$.

Consider a singular kernel $K(x) = \dfrac{\Omega(x')}{|x|^n}$, $x \neq 0$, $x = |x| x'$, and suppose that Ω is integrable on the unit sphere Σ and has mean value zero, i.e.

$$\int_\Sigma \Omega(x')\, dx' = 0.$$

If $0 < \varepsilon < \eta < \infty$, we form the truncated kernel

$$K_{\varepsilon,\eta}(x) = \begin{cases} K(x) & \text{for } \varepsilon < |x| < \eta \\ 0 & \text{otherwise,} \end{cases}$$

and we set $\tilde{f}_{\varepsilon,\eta} = f*K_{\varepsilon,\eta}$. Now $K_{\varepsilon,\eta} \in L^1(E^n)$, so if $f \in L^2$ we have that $\tilde{f}_{\varepsilon,\eta} \in L^2$ and has Fourier transform $\hat{\tilde{f}}_{\varepsilon,\eta} = \hat{f}\,\hat{K}_{\varepsilon,\eta}$.

LEMMA 3: If Ω is bounded, then for some constant $A > 0$

(i) $|\hat{K}_{\varepsilon,\eta}(x)| \leq A$, where A is independent of x, ε and η.

(ii) There exists a function $\hat{K}$ such that $\hat{K}_{\varepsilon,\eta}(x) \to \hat{K}(x)$, for all $x \neq 0$, as $\varepsilon \to 0$ and $\eta \to \infty$ simultaneously or successively.

Proof: Let $r = |x|$, $p = |y|$, $(x\cdot y) = rp\cos\theta$, and $dy = p^{n-1}dp\,dy'$. Then,

$$\hat{K}_{\varepsilon,\eta}(x) = \int_{\varepsilon<|y|<\eta} K(y)e^{-2\pi i(x\cdot y)}dy = \int_{\Sigma} \Omega(y')\left[\int_{\varepsilon}^{\eta} e^{-2\pi irp\cos\theta}\frac{dp}{p}\right]dy'$$

If $x \neq 0$, so $r > 0$, and we let $s = rp$, we have

$$\hat{K}_{\varepsilon,\eta}(x) = \int_{\Sigma} \Omega(y')\left[\int_{\varepsilon r}^{\eta r} e^{-2\pi is\cos\theta}\frac{ds}{s}\right]dy'.$$

As $\varepsilon \to 0$, the inner integrand behaves like $\frac{1}{s}$ which is not integrable. But, since Ω has mean value zero on Σ, we can correct this situation by the usual device of subtracting some function $g(s)$. Many possible choices of g will do; the simplest one is probably to take $g(s)$ to be the characteristic function of $(0,1)$.

Let $g(s) = 1$ if $0 < s < 1$ and $g(s) = 0$ elsewhere. Then

$$\hat{K}_{\varepsilon,\eta}(x) = \int_{\Sigma} \Omega(y')\left[\int_{\varepsilon r}^{\eta r} \frac{e^{-2\pi is\cos\theta} - g(s)}{s}\,ds\right]dy'$$

and we examine the inner integral

$$I = \int_{\varepsilon r}^{\eta r} [\exp(-2\pi is\cos\theta) - g(s)]\frac{ds}{s}.$$

CLAIM: For $\cos\theta \neq 0$, and some positive constant C, we have the estimate

$$|I| \leq 2\log\frac{1}{|\cos\theta|} + C.$$

We may assume that $\cos\theta > 0$, since otherwise we can estimate the complex conjugate of I. Moreover, we note that the integral

$$\int_{1}^{\infty} \frac{e^{-iu}}{u}\,du \quad \text{is convergent.}$$

In fact an integration by parts gives

$$\int_1^\infty \frac{e^{-iu}}{u}\,du = -ie^{-i} + i\int_1^\infty \frac{e^{-iu}}{u^2}\,du$$

where the last integral is absolutely convergent. We then let

$$C_1 = \sup_v \left| \int_1^v \frac{e^{-iu}}{u}\,du \right| .$$

To estimate I we distinguish three cases:

1) $\varepsilon r \leqslant 1 \leqslant \eta r$

2) $\varepsilon r > 1$

3) $\eta r < 1$.

Case 1): Since $g(s)$ is the characteristic function of $(0,1)$, we have

$$I = \int_{\varepsilon r}^{1} \frac{\exp(-2\pi is\cos\theta)-1}{s}\,ds + \int_1^{\eta r} \exp(-2\pi is\cos\theta)\,\frac{ds}{s} = I_1 + I_2.$$

Since $|e^{iu}-1| = |e^{iu/2}(e^{iu/2}-e^{-iu/2})| \leqslant |2i\sin\frac{u}{2}| \leqslant |u|$, it follows that

$$|I_1| \leqslant \int_{\varepsilon r}^{1} \frac{|2\pi s\cos\theta|}{s}\,ds \leqslant 2\pi.$$

For I_2, we let $t = s\cos\theta$ so that $I_2 = \int_{\cos\theta}^{\eta r\cos\theta} e^{-2\pi it}\,\frac{dt}{t}$.

If $\eta r\cos\theta > 1$, then

$$|I_2| \leqslant \int_{\cos\theta}^{1} \frac{dt}{t} + \left| \int_1^{\eta r\cos\theta} e^{-2\pi it}\,\frac{dt}{t} \right| \leqslant \log\frac{1}{|\cos\theta|} + C_1.$$

On the other hand, if $0 < \eta r\cos\theta \leqslant 1$, then $|I_2| \leqslant \int_{\cos\theta}^{1} \frac{dt}{t} = \log\frac{1}{|\cos\theta|}$

Therefore, combining estimates, we obtain

$$|I| \leqslant \log\frac{1}{|\cos\theta|} + C_1 + 2\pi.$$

Case 2): Here $\varepsilon r > 1$, so $g(s) = 0$ and

$$I = \int_{\varepsilon r}^{\eta r} \exp(-2\pi is\cos\theta)\,\frac{ds}{s} = \int_1^{\eta r} - \int_1^{\varepsilon r} = I_2 - I_3.$$

As we saw in the previous case, $|I_2| \leq \log \frac{1}{|\cos\theta|} + C_1$. Repeating that argument we obtain the same estimate for $|I_3|$. Hence,

$$|I| \leq 2 \log \frac{1}{|\cos\theta|} + 2C_1 .$$

Case 3): Since $\eta r < 1$, we have, as in Case 1), that

$$|I| \leq \int_0^1 \frac{|2\pi s \cos\theta|}{s} \, ds \leq 2\pi .$$

Therefore, taking $C = 2C_1 + 2\pi$, we have that in all cases

$$|I| \leq 2 \log \frac{1}{|\cos\theta|} + C$$

and the Claim is proved.

Since I is majorized by a function which is integrable on Σ then, if Ω is bounded, it follows that

$$(*) \qquad |\hat{K}_{\varepsilon,\eta}(x)| \leq \int_\Sigma |\Omega(y')| [2 \log \frac{1}{|\cos\theta|} + C] \, dy' = A < \infty$$

so part (i) is proved.

The second part follows from the fact that I converges a.e. as $\varepsilon \to 0$ and $\eta \to \infty$, and that it is dominated by a function integrable on Σ. Therefore, by Lebesgue's Theorem of Dominated Convergence, we conclude that, for $x \neq 0$, $\hat{K}_{\varepsilon,\eta}(x)$ converges to a function $\hat{K}(x)$.

An explicit formula for $\hat{K}$ will be given later.

Q.E.D.

We observe from estimate (*) above that for the validity of Lemma 3 we only need, besides the mean value zero condition, that

$$(**) \qquad \int_\Sigma |\Omega(y')| \log \frac{1}{|\cos\theta|} \, dy' < \infty .$$

This will be true, by Hölder's inequality, if for instance $\Omega \in L^r(\Sigma)$ for some $r > 1$. However, we can obtain a slightly more general condition on Ω in the following way.

First of all, (**) implies that Ω must be integrable on Σ. If u and v are strictly increasing, continuous functions on $[o,\infty)$ such that $v = u^{-1}$ and $u(o) = v(o) = 0$, then for any $a, b \geqslant 0$ we have Young's inequality:

$$ab \leqslant \int_0^a u(s)\, ds + \int_0^b v(t)\, dt.$$

Choosing $t = u(s) = \log(1+s)$, so $s = v(t) = e^t - 1$, we obtain

$$ab \leqslant a \log(1+a) + e^b - 1 - b < a \log(1+a) + e^b.$$

Letting $a = |\Omega(y')|$ and $b = \lambda \log \frac{1}{|\cos\theta|}$, where $0 < \lambda < 1$, we have that

$$|\Omega(y')| \log \frac{1}{|\cos\theta|} \leqslant \frac{1}{\lambda} |\Omega(y')| \log(1+|\Omega(y')|) + \frac{1}{\lambda} \frac{1}{|\cos\theta|^\lambda}.$$

Integrating over Σ, we note that since $\lambda < 1$

$$\int_\Sigma \frac{dy'}{|\cos\theta|^\lambda} < \infty$$

therefore (**) holds provided that $|\Omega| \log(1+|\Omega|) \in L^1(\Sigma)$, or equivalently, that $|\Omega| \log^+|\Omega| \in L^1(\Sigma)$. As we mentioned at the end of last section, this condition is expressed by saying that $\Omega \in L \log^+ L(\Sigma)$.

We can now improve the preceding Lemma as follows.

LEMMA 3': If $\Omega \in L \log^+ L(\Sigma)$ and Ω has mean value zero on Σ, then $\hat{K}_{\varepsilon,\eta}(x)$ is bounded in x, uniformly in ε and η, and there exists a function $\hat{K}$ such that, for $x \neq 0$, $\hat{K}_{\varepsilon,\eta}(x) \to \hat{K}(x)$ as $\varepsilon \to 0$ and $\eta \to \infty$.

THEOREM 3: (Calderòn-Zygmund) Let $f \in L^2(E^n)$ and suppose that $\Omega \in L \log^+(\Sigma)$ and Ω has mean value zero on Σ. Set $\tilde{f}_{\varepsilon,\eta} = f * K_{\varepsilon,\eta}$. Then,

a) $\|\tilde{f}_{\varepsilon,\eta}\|_2 \leqslant A\|f\|_2$, where $A > 0$ is independent of ε, η and f.

b) There exists a function $\tilde{f} \in L^2$ such that $\|\tilde{f}_{\varepsilon,\eta} - \tilde{f}\|_2 \to 0$ as $\varepsilon \to 0$ and $\eta \to \infty$.

c) $\hat{\tilde{f}} = \hat{f}\hat{K}$ and $\|\tilde{f}\|_2 \leqslant M\|f\|_2$, where $M = \sup_x |\hat{K}(x)|$.

d) For $f \in \Lambda^{\alpha} \cap L^2$, the pointwise limit $(Kf)(x)$ equals $\tilde{f}(x)$ a.e..

Proof: By Lemma 3', $|\hat{K}_{\varepsilon,\eta}(x)| \leq A$. So, using Parseval's formula,

$$\|\tilde{f}_{\varepsilon,\eta}\|_2 = \|(\tilde{f}_{\varepsilon,\eta})^{\wedge}\|_2 = \|\hat{f}\,\hat{K}_{\varepsilon,\eta}\|_2 \leq A\|\hat{f}\|_2 = A\|f\|_2$$

and a) is proved.

Define the function $\tilde{f} \in L^2$ by the formula $\hat{\tilde{f}} = \hat{f}\,\hat{K}$. Then part c) follows at once. Part d) follows directly from b) and Theorem 1.

Finally, as $\varepsilon \to 0$ and $\eta \to \infty$,

$$\|\tilde{f}_{\varepsilon,\eta} - \tilde{f}\|_2 = \|(\tilde{f}_{\varepsilon,\eta} - \tilde{f})^{\wedge}\|_2 = \|(\hat{K}_{\varepsilon,\eta} - \hat{K})\hat{f}\|_2 \to 0$$

by Lebesgue's Theorem of Dominated Convergence, since $|\hat{K}_{\varepsilon,\eta} - \hat{K}| \leq A+M$ and $|\hat{K}_{\varepsilon,\eta}(x) - \hat{K}(x)| \to 0$, for $x \neq 0$.

Q.E.D.

THEOREM 4: Let $\Omega \in L\log^{+}L(\Sigma)$, $\int_{\Sigma} \Omega(x')\,dx' = 0$, and $K(x) = \dfrac{\Omega(x')}{|x|^n}$, $x \neq 0$. Then

$$(1)\quad \hat{K}(x) = \int_{\Sigma} \Omega(y') \left[\log \frac{1}{|\cos\theta|} - \frac{i\pi}{2}\,\operatorname{sgn}(\cos\theta) \right] dy'$$

where θ is the angle between x and y.

Proof: In the proof of Lemma 3 we saw that if $r = |x| > 0$ and $s = r|y|$, then

$$\hat{K}_{\varepsilon,\eta}(x) = \int_{\Sigma} \Omega(y') \left[\int_{\varepsilon r}^{\eta r} e^{-2\pi i s\cos\theta}\,\frac{ds}{s} \right] dy'$$

where the inner integral is convergent as $\eta \to \infty$. Therefore, for $x \neq 0$, $\hat{K}$ is defined by the formula

$$\hat{K}(x) = \lim_{\delta \to 0} \int_{\Sigma} \Omega(y') \left[\int_{\delta}^{\infty} e^{-2\pi i s\cos\theta}\,\frac{ds}{s} \right] dy'.$$

Expressing the inner integral I in terms of its real and imaginary parts, $I = U + iV$, and letting $t = 2\pi s \cos\theta$, we see that

$$\lim_{\delta\to 0} V = -\operatorname{sgn}(\cos\theta)\int_0^\infty \frac{\sin t}{t}\,dt = -\frac{\pi}{2}\operatorname{sgn}(\cos\theta).$$

On the other hand, we can write U in the form

$$U = \int_{2\pi\delta|\cos\theta|}^\infty \frac{\cos t}{t}\,dt = \int_1^\infty \frac{\cos t}{t}\,dt + \int_{2\pi\delta|\cos\theta|}^1 \frac{\cos t-1}{t}\,dt +$$

$$+ \int_{2\pi\delta|\cos\theta|}^1 \frac{dt}{t} = C + \int_{2\pi\delta|\cos\theta|}^1 \frac{\cos t-1}{t}\,dt + \log\frac{1}{|\cos\theta|} + C(\delta)$$

and since Ω has mean value zero on Σ we obtain

$$\int_\Sigma \Omega(y')U\,dy' = \int_\Sigma \Omega(y')\left[\int_{2\pi\delta|\cos\theta|}^1 \frac{\cos t-1}{t}\,dt\right]dy' +$$

$$+ \int_\Sigma \Omega(y')\log\frac{1}{|\cos\theta|}\,dy'.$$

As we saw earlier, the last integral is finite for $\Omega \in L\log^+L(\Sigma)$. In particular, $\Omega \in L^1(\Sigma)$ and since in the first term the inner integral converges boundedly as $\delta \to 0$, we can pass the limit under the outer integral sign and conclude, in view of the mean value zero condition, that

$$\hat{K}(x) = \int_\Sigma \Omega(y')\lim_{\delta\to 0}[U+iv]\,dy' = \int_\Sigma \Omega(y')\left[\log\frac{1}{|\cos\theta|} - \frac{i\pi}{2}\operatorname{sgn}(\cos\theta)\right]dy'.$$

Q.E.D.

We observe that $\hat{K}$ has the following properties:

a) $\hat{K}$ is bounded.

This follows immediately from Lemma 3'.

b) $\hat{K}$ is homogeneous of degree zero.

This can be seen directly from formula (1). More generally,

if a function f on E^n is homogeneous of degree $-\alpha$, then $\hat{f}$ is homogeneous of degree $-n+\alpha$. In fact, formally, if $t = \lambda y$,

$$\hat{f}(\lambda x) = \int f(y) e^{-2\pi i(\lambda x\cdot y)} dy = \int \lambda^{\alpha} f(t) e^{-2\pi i(x.t)} \frac{dt}{\lambda^n} = \lambda^{-n+\alpha}\hat{f}(x).$$

c) $\hat{K}$ has mean value zero on Σ.

In fact, interchanging the order of integration,

$$\int_\Sigma \hat{K}(x)\,dx' = \int_\Sigma dx' \int_\Sigma \Omega(y') \left[\log \frac{1}{|\cos\theta|} - \frac{i\pi}{2}\operatorname{sgn}(\cos\theta)\right] dy' =$$

$$= \int_\Sigma \Omega(y') \left[\int_\Sigma \log \frac{1}{|\cos\theta|} - \frac{i\pi}{2}\operatorname{sgn}(\cos\theta)\,dx'\right] dy' =$$

$$= C\int_\Sigma \Omega(y')\,dy' = 0.$$

d) If K is odd, then

$$\hat{K}(x) = -\frac{\pi i}{2}\int_\Sigma \Omega(y')\operatorname{sgn}(\cos\theta)\,dy' = -\pi i \int_{\Sigma^+(x)} \Omega(y')\,dy'$$

where $\Sigma^+(x)$ denotes the hemisphere where $\cos\theta \geqslant 0$.

Summing up, we have proved that if $K(x) = \Omega(x')|x|^{-n}$ is a kernel homogeneous of degree $-n$, where $\Omega \in L\log^+L(\Sigma)$ and has mean value zero on Σ, the singular integral operator

$$\tilde{f}(x) = \text{p.v.}(f*K)(x) = \lim_{\varepsilon\to 0}\int_{|y|>\varepsilon} f(x-y)K(y)\,dy$$

exists as a limit in L^2 norm and is a bounded linear transformation of $L^2(E^n)$. Moreover, since $\hat{\tilde{f}} = \hat{K}\hat{f}$, where $\hat{K}$ is given by formula (1) of Th. 4, we can write the operator $Hf = \tilde{f}$ in the form

$$H = F^{-1}\hat{K}F$$

where F and F^{-1} denote respectively the Fourier and Inverse Fourier transform. In view of Exercise 2 at the end of last chapter, $H: L^2 \to L^2$ has norm

$$\|H\| = \sup_x |\hat{K}(x)|$$

so the constant M of Th. 3 is the best possible.

We conclude this section with a result on the Riesz transforms which will be used later.

Consider the Riesz transforms

$$R_j f = \text{p.v.}\ (f * \frac{x_j}{|x|^{n+1}}) \qquad j = 1,2,\ldots,n.$$

LEMMA 4: If $K(x) = \frac{x_j}{|x|^{n+1}}$, then $\hat{K}(x) = \gamma \frac{x_j}{|x|}$, where γ is a constant depending only on n.

Proof: Since K(x) is an odd kernel then, by property d) above,

$$\hat{K}(x) = -\pi i \int_{\Sigma^+(x)} \Omega(y')\, dy'.$$

Clearly it suffices to consider the case j = 1. Here $\Omega(y') = y_1' = y' \cdot x_1$. Relative to any other orthonormal basis $\{z_1,\ldots,z_n\}$, we have that $y' = \Sigma_j (y' \cdot z_j) z_j$ and hence

$$\Omega(y') = y' \cdot x_1 = \sum_{j=1}^{n} (y' \cdot z_j)(z_j \cdot x_1).$$

Fixing $x \in E^n$, $x = |x|x'$, let us choose an orthonormal basis $\{z_1,\ldots,z_n\}$ with $z_1 = x'$. Then $z_1 \cdot x_1 = x' \cdot x_1 = x_1'$, and letting $c_j = z_j \cdot x_1$ for $j = 2,\ldots,n$, we have that

$$\Omega(y') = x_1'(y' \cdot x') + \sum_{j=2}^{n} c_j (y' \cdot z_j).$$

Integrating over our hemisphere where $y' \cdot x' \geq 0$, we see that

$$\hat{K}(x) = -\pi i x_1' \int_{\Sigma^+(x)} (y' \cdot x')\, dy' \ -\pi i \sum_{j=2}^{n} c_j \int_{\Sigma^+(x)} (y' \cdot z_j) dy'.$$

Now, the first integral is equal to v_{n-1}, the volume of the unit ball in E^{n-1}, whereas the integrals in the second term vanish. Therefore,

$$\hat{K}(x) = \gamma x_1' = \gamma \frac{x_1}{|x|}, \text{ where } \gamma = -\pi i v_{n-1}.$$

Q.E.D.

REMARK: (Calculation of γ)

Denote by v_n the volume of the unit ball in E^n and by w_n the area of the unit sphere in E^n. Recall the definition of the gamma function:

$$\Gamma(s) = \int_0^\infty e^{-u} u^{s-1} du.$$

The volume of a ball of radius r in E^n is

$$v_n r^n = \int_0^r w_n p^{n-1} dp = w_n \frac{r^n}{n}$$

so

(i) $$w_n = n v_n.$$

Now,

$$I = \int e^{-|x|^2} dx = \left[\int_{-\infty}^{\infty} e^{-t^2} dt \right]^n = (\sqrt{\pi})^n$$

and, in polar coordinates,

$$I = \int_\Sigma dx' \int_0^\infty e^{-r^2} r^{n-1} dr = \frac{1}{2} w_n \int_0^\infty e^{-u} u^{\frac{n}{2}-1} du = \frac{1}{2} w_n \Gamma(\tfrac{n}{2})$$

whence

(ii) $$w_n = \frac{2(\sqrt{\pi})^n}{\Gamma(\frac{n}{2})}.$$

From (i) and (ii),

$$v_n = \frac{(\sqrt{\pi})^n}{\frac{n}{2}\Gamma(\frac{n}{2})} = \frac{(\sqrt{\pi})^n}{\Gamma(1+\frac{n}{2})}$$

so

so $v_{n-1} = \dfrac{(\sqrt{\pi})^{n-1}}{\Gamma(\frac{n+1}{2})}$ and finally

$$\gamma = -\pi i v_{n-1} = -i \frac{(\sqrt{\pi})^{n+1}}{\Gamma(\frac{n+1}{2})}.$$

In view of Lemma 4, if we normalize the Riesz transforms by redefining

$$(R_j f)(x) = \frac{1}{\gamma}\ \text{p.v.} \int f(x-y)\, \frac{y_j}{|y|^{n+1}}\, dy, \qquad j = 1,\dots,n$$

we deduce the following result due to Horvàth.

THEOREM 5: Let I denote the identity operator in $L^p(E^n)$, $1 < p < \infty$. Then

$$\sum_{j=1}^{n} R_j^2 = I.$$

Proof: For $f \in L^2$ and $j = 1,\dots,n$, $(R_j f)\hat{}\,(x) = \frac{x_j}{|x|}\hat{f}(x)$; hence

$$\sum_{j=1}^{n} (R_j^2 f)\hat{}\,(x) = \sum_{j=1}^{n} \left(\frac{x_j}{|x|}\right)^2 \hat{f}(x) = \hat{f}(x).$$

Taking Inverse Fourier transform,

$$\sum_{j=1}^{n} R_j^2 f = f$$

so the formula is valid in L^2. In particular it is valid on C_o^1 which is dense in the spaces L^p.

Since, by Th. 2 in §1, the operators R_j are continuous in L^p, the formula holds, by continuity, in every L^p, with $1 < p < \infty$.

Q.E.D.

§3. EVEN KERNELS. CONCLUSIONS

We seek the analogue of Th. 2, mentioned in the first remark at the end of §1. Denoting by the same letter a kernel and the corresponding operator, we consider a singular integral operator $Kf = $ $= \text{p.v.}\ (K*f)$, where $K(x) = \Omega(x')\ |x|^{-n}$ is an even function, Ω has mean value zero on Σ, and we assume, for simplicity, that $\Omega \in L^q(\Sigma)$ for some $q > 1$.

We saw that the (normalized) Riesz transforms

$$R_j f = \text{p.v.}(R_j * f) = \frac{1}{\gamma}\ \text{p.v.} \left(f * \frac{x_j}{|x|^{n+1}} \right)$$

are operators with odd kernels such that $\sum_{j=1}^{n} R_j^2 f = f$. Hence, formally,

$$Kf = \sum_{j=1}^{n} R_j^2 Kf = \sum_{j=1}^{n} [R_j(R_j K)]f$$

where

$$R_j Kf = \frac{1}{\gamma} \frac{x_j}{|x|^{n+1}} * (K(x)*f) = \frac{1}{\gamma} \left(\frac{x_j}{|x|^{n+1}} * K(x) \right) *f$$

and it will follow from Lemma 5 below that this last kernel $R_j K$ is odd and homogeneous of degree -n. Therefore, a singular integral operator K with even kernel can be expressed as a finite sum of products of operators with odd kernels.

Suppose now that K(x) satisfies the assumptions above and let $\tilde{f}_\varepsilon = f*K_\varepsilon$, where K_ε is the truncated kernel: $K_\varepsilon(x) = K(x)$ if $|x| > \varepsilon$, $K_\varepsilon(x) = 0$ otherwise. If $f \in L^p(E^n)$, $1 < p < \infty$, our main objective will be to prove an estimate of the form

$$\|\tilde{f}_\varepsilon\|_p \leqslant B_{p,q} \|f\|_p$$

for some positive constant $B_{p,q}$ independent of ε and f. The idea of the proof will be to express the operator $K_\varepsilon f = f*K_\varepsilon$ in the manner indicated in the preceding paragraph.

To begin with, we need the following Lemmas.

LEMMA 5: Let $h(x) = (f*g)(x)$, $x \in E^n$. Then h is even if f and g have the same parity, and h is odd if f and g have opposite parity. If f and g are homogeneous of degrees α_1 and α_2 respectively, then h is homogeneous of degree $\alpha_1 + \alpha_2 + n$.

Proof: Since

$$h(x) = \int f(y)g(x-y)\,dy$$

and

$$h(-x) = \int f(y)g(-x-y)\,dy = \int f(-y)g(-x+y)\,dy$$

the first conclusion follows directly.

Moreover, if $\lambda > 0$ and we let $y = \lambda z$, we obtain

$$h(\lambda x) = \int f(y)g(\lambda x-y)\,dy = \int f(\lambda z)g(\lambda x-\lambda z)\lambda^n\,dz =$$
$$= \int \lambda^{\alpha_1} f(z)\lambda^{\alpha_2} g(x-z)\lambda^n\,dz = \lambda^{\alpha_1+\alpha_2+n} h(x).$$

Q.E.D.

LEMMA 6: Let $f \in L^1(E^n)$, $\Omega \in L^q(\Sigma)$, $1 < q < \infty$. Then,

$$R_j(K_\varepsilon f) = (R_j K_\varepsilon)*f.$$

Proof: First of all, $K_\varepsilon \in L^q$, since with $r = |x|$

$$\int |K_\varepsilon(x)|^q dx = \int_{|x|>\varepsilon} |K(x)|^q dx = \int_\Sigma |\Omega(x')|^q dx' \int_\varepsilon^\infty \frac{dr}{r^{1+n(q-1)}} = C\|\Omega\|_q^q < \infty.$$

Thus $K_\varepsilon f = (K_\varepsilon * f) \in L^q$, since $f \in L^1$. By Th. 2 then, $R_j(K_\varepsilon f) \in L^q$ and $R_j K_\varepsilon = \text{p.v.}(R_j * K_\varepsilon) \in L^q$, therefore $(R_j K_\varepsilon)*f$ is also in L^q.

Again by Th. 2, $R_j(K_\varepsilon f) = \lim_{\delta\to 0} R_{j,\delta}(K_\varepsilon f)$, where the limit is taken in the L^q norm and $R_{j,\delta}$ is the operator given by the Riesz kernel R_j truncated at δ. Moreover,

$$[R_{j,\delta}(K_\varepsilon f)](x) = \int R_{j,\delta}(x-y)\left[\int f(z)K_\varepsilon(y-z)dz\right]dy =$$

$$= \int f(z)\left[\int R_{j,\delta}(x-y)K_\varepsilon(y-z)dy\right]dz$$

where the interchange of the order of integration is justified because the last integral is absolutely convergent: in fact f is integrable and the inner integral is bounded in absolute value by $\|K_\varepsilon\|_q \|R_{j,\delta}\|_{q'}$ where $q' = q/(q-1)$, as can be seen from Hölder's inequality. Hence, letting $y = z + t$,

$$[R_{j,\delta}(K_\varepsilon f)](x) = \int f(z)\left[\int R_{j,\delta}(x-z-t)K_\varepsilon(t)dt\right]dz =$$

$$= \int f(z)[R_{j,\delta}*K_\varepsilon](x-z)dz = [f*(R_{j,\delta}*K_\varepsilon)](x).$$

Since $K_\varepsilon \in L^q$, we have by Th. 2 that $(R_{j,\delta}*K_\varepsilon) \to (R_j*K_\varepsilon)$ in L^q norm, as $\delta \to 0$. By Young's theorem, convolution with $f \in L^1$ is a continuous operation in L^q. Hence, as $\delta \to 0$,

$$f*(R_{j,\delta}*K_\varepsilon) \to f*(R_j*K_\varepsilon)$$

in L^q norm. In other words,

$$\lim_{\delta\to 0} \;(\text{in } L^q)\; R_{j,\delta}(K_\varepsilon f) = f*(R_j*K_\varepsilon) =$$

$$= (R_j*K_\varepsilon)*f = (R_jK_\varepsilon)*f.$$

Therefore $R_j(K_\varepsilon f) = (R_jK_\varepsilon)*f$.

Q.E.D.

LEMMA 7: Let $K(x) = \Omega(x')|x|^{-n}$ be even and homogeneous of degree $-n$, $\Omega \in L^q(\Sigma)$, $q \geq 1$, and with mean value zero on Σ. Suppose that $j \in [1,n]$ is a fixed integer.

Then, there exists a kernel $\tilde{K}$ which is odd, homogeneous of degree $-n$, and such that, as $\varepsilon \to 0$, $R_j*K_\varepsilon \to \tilde{K}$ in the L^∞ metric over any closed subset of E^n which does not contain the origin.

Proof: If $0 < \varepsilon < \eta < \infty$, then

$$[R_j*(K_\varepsilon-K_\eta)](x) = \int R_j(x-y)[K_\varepsilon-K_\eta](y)dy =$$

$$= \int_{\varepsilon<|y|<\eta} R_j(x-y)K(y)dy.$$

Since Ω has mean value zero on Σ, $K(y)$ has mean value zero on $\varepsilon<|y|<\eta$, so

$$[R_j*(K_\varepsilon-K_\eta)](x) = \int_{\varepsilon<|y|<\eta} [R_j(x-y)-R_j(x)]K(y)dy.$$

Now, using the Mean Value Theorem, for some $0 < \theta < 1$,

$$\left|R_j(x-y)-R_j(x)\right| \leqslant \left|\frac{x_j-y_j}{|x-y|^{n+1}} - \frac{x_j}{|x|^{n+1}}\right| \leqslant \frac{(n+2)|y|}{|x-\theta y|^{n+1}} .$$

If we choose $\varepsilon < |y| < \eta < \frac{|x|}{2}$, then $|x-\theta y| > |x| - |y| > \frac{|x|}{2}$ and so

$$|R_j(x-y) - R_j(x)| \leqslant \frac{C}{|x|^{n+1}} |y| .$$

Therefore, it follows that

$$|[R_j*(K_\varepsilon-K_\eta)](x)| \leqslant C|x|^{-n-1} \int_{\varepsilon<|y|<\eta} |y|\ |K(y)|dy =$$

$$= C|x|^{-n-1} \int_\varepsilon^\eta dr \int_\Sigma |\Omega(y')|dy' \leqslant$$

$$\leqslant C|x|^{-n-1} \|\Omega\|_1 \eta \to 0$$

as $\eta \to 0$ and x is bounded away from the origin. In other words $\{R_j*K_\varepsilon\}$ is a Cauchy sequence in the L^∞ norm, over all closed subsets which do not contain the origin.

For any $\alpha > 0$, we denote by $K^*(x)$ the pointwise limit, as $\varepsilon \to 0$, of $(R_j*K_\varepsilon)(x)$ on $|x| \geqslant \alpha$. Since, by Lemma 5, R_j*K_ε is odd, the function K^* will be odd almost everywhere. Therefore, correcting the function on a set of measure zero, we may assume that K^* is odd everywhere.

Moreover, for all x and all $\lambda > 0$, we obtain, letting $y = \lambda z$,

$$(R_{j,\delta}K_\varepsilon)(\lambda x) = \int R_{j,\delta}(\lambda x - y)K_\varepsilon(y)\,dy = \int R_{j,\delta}(\lambda[x-z])K_\varepsilon(\lambda z)\lambda^n dz =$$

$$= \int \lambda^{-n}R_{j,\delta/\lambda}(x-z)\lambda^{-n}K_{\varepsilon/\lambda}(z)\lambda^n dz =$$

$$= \lambda^{-n}(R_{j,\delta/\lambda}K_{\varepsilon/\lambda})(x).$$

Hence, for any fixed $\lambda > 0$, it follows that for almost every x

$$K^*(\lambda x) = \lambda^{-n}K^*(x).$$

Here the exceptional set of points x where the equality does not hold depends on λ. However, since K^* is measurable, the equality will be valid for almost every point (x,λ) in $E^n \times (0,\infty)$.

Denote by Z the set of points x for which the preceding equality is violated on a set of values of λ having positive measure:

$$Z = \{x: |\{\lambda: K^*(\lambda x) \neq \lambda^{-n}K^*(x)\}| > 0\}.$$

Clearly, Z must have measure zero. But then it follows from Fubini's theorem that for some $\varrho > 0$, the intersection of the sphere $\Sigma_\varrho = \{x: |x| = \varrho\}$ with the set Z has zero surface measure. With this ϱ we define

$$\tilde{K}(x) = \begin{cases} \left(\frac{\varrho}{|x|}\right)^n K^*\left(\varrho \frac{x}{|x|}\right) & \text{if } x \neq 0 \text{ and } \varrho \frac{x}{|x|} \notin Z \\ 0 & \text{otherwise.} \end{cases}$$

Now, $\tilde{K}$ is measurable, odd, and homogeneous of degree -n. It remains to show that $\tilde{K} = K^*$ almost everywhere.

Let $x \neq 0$ so $x_o = \varrho \frac{x}{|x|} \in \Sigma_\varrho$. If $x_o \notin Z$, then $\tilde{K}(x_o) = K^*(x_o)$ and so, since $\Sigma_\varrho \cap Z$ has zero surface measure, $\tilde{K} = K^*$ almost everywhere on Σ_ϱ. Moreover, $\tilde{K}(\lambda x_o) = \lambda^{-n}\tilde{K}(x_o) = \lambda^{-n}K^*(x_o)$ and, since $\lambda^{-n}K^*(x_o) = K^*(\lambda x_o)$ for almost every $\lambda > 0$, it follows that $\tilde{K}(x) = K^*(x)$ almost everywhere.

Q.E.D.

Lemma 8: Let K and $\tilde{K}$ be as in Lemma 7. If $1 < q < \infty$, then

(i) $\int_{\Sigma} |\tilde{K}(x')|\,dx' \leq A_q \|\Omega\|_q.$

If we set $\Delta_\varepsilon = R_j K_\varepsilon - (\tilde{K})_\varepsilon$, where $(\tilde{K})_\varepsilon$ denotes the truncation of $\tilde{K}$ at ε, then Δ_ε is integrable and

(ii) $\|\Delta_\varepsilon\|_1 \leq A'_q \|\Omega\|_q.$

Proof: Since $\tilde{K}$ is homogeneous of degree $-n$ we observe, using polar coordinates, that

$$1)\quad \int_{\Sigma} |\tilde{K}(x')|\,dx' = (\log 2)^{-1} \int_{1<|x|<2} |\tilde{K}(x)|\,dx.$$

If $|x| \geq 1$ then, by an argument used in the previous Lemma, choosing $\eta = 1/2$ and letting $\varepsilon \to 0$ we obtain

$$|R_j K_{1/2}(x) - \tilde{K}(x)| \leq \frac{C}{|x|^{n+1}} \|\Omega\|_1 \leq \frac{C_q}{|x|^{n+1}} \|\Omega\|_q$$

whence

$$2)\quad \int_{1<|x|<2} |R_j K_{1/2}(x) - \tilde{K}(x)|\,dx \leq C_q \|\Omega\|_q \int_{|x|>1} \frac{dx}{|x|^{n+1}} = C'_q \|\Omega\|_q.$$

Since $q > 1$, we know that $K_{1/2} \in L^q$, in fact $\|K_{1/2}\|_q = B\|\Omega\|_q$. Moreover, by Th. 2, $\|R_j K_{1/2}\|_q \leq B_q \|K_{1/2}\|_q$ for $1 < q < \infty$. Therefore, using Hölder's inequality, we have that

$$3)\quad \int_{1<|x|<2} |R_j K_{1/2}(x)|\,dx \leq A\|R_j K_{1/2}\|_q \leq B'_q \|\Omega\|_q.$$

But $|\tilde{K}(x)| \leq |\tilde{K}(x) - R_j K_{1/2}(x)| + |R_j K_{1/2}(x)|$ so, combining 1), 2) and 3), we obtain (i).

Next, a simple calculation shows that $\Delta_\varepsilon(x) = \varepsilon^{-n} \Delta_1(x/\varepsilon)$; hence Δ_ε and Δ_1 have the same L^1 norm. Now,

$$\|\Delta_\varepsilon\|_1 = \|\Delta_1\|_1 = \int |R_j K_1(x) - (\tilde{K})_1(x)|\,dx \leq$$

$$\leqslant \int_{|x|<2} |R_jK_1(x)|\,dx + \int_{1<|x|<2} |\tilde{K}(x)|\,dx + \int_{|x|\geqslant 2} |\Delta_1(x)|\,dx =$$

$$= I_1 + I_2 + I_3.$$

As in estimate 3), we have that

$$I_1 \leqslant A\|R_jK_1\|_q \leqslant A_q\|K_1\|_q = A_q\|K_1\|_q = A_q'\|\Omega\|_q.$$

By 1) and (i), we see that

$$I_2 = (\log 2)\int_\Sigma |\tilde{K}(x')|\,dx' \leqslant A_q'\|\Omega\|_q.$$

Finally, as in estimate 2), we have that

$$I_3 \leqslant A_q\|\Omega\|_q \int_{|x|\geqslant 2} \frac{dx}{|x|^{n+1}} = A_q'\|\Omega\|_q.$$

The constants A_q' are of course all different. Adding up, we obtain (ii).

Q.E.D.

We are now in a position to prove the following theorem.

THEOREM 6: Suppose that $K(x) = \Omega(x')|x|^{-n}$ is an even kernel with mean value zero on Σ and that $\Omega \in L^q(\Sigma)$ for some $q > 1$. If $f \in L^p(E^n)$, $1 < p < \infty$, we set $\tilde{f}_\varepsilon = K_\varepsilon f = K_\varepsilon * f$, where K_ε is the truncated kernel: $K_\varepsilon(x) = K(x)$ if $|x| > \varepsilon$, $K_\varepsilon(x) = 0$ otherwise.

Then,

a) $\|\tilde{f}_\varepsilon\|_p \leqslant A_{p,q}\|\Omega\|_q \|f\|_p$.

b) There exists a function $\tilde{f} \in L^p$ such that $\|\tilde{f}_\varepsilon - \tilde{f}\|_p \to 0$ as $\varepsilon \to 0$.

c) $\|\tilde{f}\|_p \leqslant A_{p,q}\|\Omega\|_q \|f\|_p$.

Proof: It suffices to prove a) for $f \in C_o^1$, and hence integrable. Then, by Lemma 6 and the definition of Δ_ε, we have that

$$R_j\tilde{f}_\varepsilon = R_j(K_\varepsilon f) = (R_jK_\varepsilon)*f = (\tilde{K})_\varepsilon * f + \Delta_\varepsilon * f$$

whence,

$$\|R_j\tilde{f}_\varepsilon\|_p \leq \|(\tilde{K})_\varepsilon * f\|_p + \|\Delta_\varepsilon * f\|_p.$$

Since $\tilde{K}$ is odd, Th. 2 and estimate (i) of Lemma 8 imply that

$$\|(\tilde{K})_\varepsilon * f\|_p \leq B_p \int_\Sigma |\tilde{K}(x')|\,dx' \;\|f\|_p \leq B_{p,q}\|\Omega\|_q \;\|f\|_p.$$

On the other hand, Young's theorem and estimate (ii) of Lemma 8 yield

$$\|\Delta_\varepsilon * f\|_p \leq \|\Delta_\varepsilon\|_1 \|f\|_p \leq C_q\|\Omega\|_q \;\|f\|_p.$$

Therefore, using Th. 5, Th. 2, and the previous estimates, we obtain

$$\|\tilde{f}_\varepsilon\|_p = \|\sum_{j=1}^{n} R_j^2 K_\varepsilon f\|_p \leq A_p \sum_{j=1}^{n} \|R_j(K_\varepsilon f)\|_p \leq A_{p,q}\|\Omega\|_q \;\|f\|_p.$$

The proof of part b) is identical to the one in Th. 2. Finally, c) is a direct consequence of a) and b).

Q.E.D.

REMARK: The preceding theorem remains valid (with the constants suitably modified) if the assumption $\Omega \in L^q(\Sigma)$ for some $q > 1$ is replaced by the weaker condition that $\Omega \in L \log^+ L(\Sigma)$. For example, we shall have

a') $\|\tilde{f}_\varepsilon\|_p \leq A_p\{\int_\Sigma |\Omega(x')|\log^+|\Omega(x')|\,dx' + A\}\|f\|_p.$

Combining Th. 2 of §1 with Th. 6 above, we deduce readily the main result of this chapter:

THEOREM 7: (Calderòn-Zygmund) Let $f \in L^p(E^n)$, $1 < p < \infty$, and set $\tilde{f}_\varepsilon = K_\varepsilon * f$ where $K(x) = \Omega(x')|x|^{-n}$ has mean value zero on Σ and $\Omega \in L^q(\Sigma)$ for some $q > 1$.

Then,

a) $\|\tilde{f}_\varepsilon\|_p \leq A_{p,q}\|\Omega\|_q \;\|f\|_p.$

b) There exists a function $\tilde{f} \in L^p$ such that $\|\tilde{f}_\varepsilon - \tilde{f}\|_p \to 0$ as $\varepsilon \to 0$.

c) $\|\tilde{f}\|_p \leqslant A_{p,q} \|\Omega\|_q \; \|f\|_p$.

Proof: As before, it is enough to verify a), and moreover we may assume that $f \in C_o^1$, for instance.

We decompose the kernel K into its even and odd parts: $K = K' + K''$, where $K'(x) = (1/2)[K(x) + K(-x)]$ and $K''(x) = (1/2)[K(x) - K(-x)]$, and we denote by Ω' and Ω'' the corresponding characteristics.

Clearly, Ω'' is odd and integrable on Σ. In particular, it has mean value zero on Σ. Hence $\Omega' = \Omega - \Omega''$ also has mean value zero on Σ. Furthermore, $\Omega' \in L^q(\Sigma)$ for some $q > 1$.

Now,

$$\|\tilde{f}_\varepsilon\|_p \leqslant \|K'_\varepsilon * f\|_p + \|K''_\varepsilon * f\|_p = I' + I''.$$

By Th. 2,

$$I'' \leqslant B_p \|\Omega''\|_1 \; \|f\|_p \leqslant B_{p,q} \|\Omega''\|_q \|f\|_p \leqslant B_{p,q} \|\Omega\|_q \; \|f\|_p.$$

By Th. 6,

$$I' \leqslant C_{p,q} \|\Omega'\|_q \; \|f\|_p \leqslant C_{p,q} \|\Omega\|_q \; \|f\|_p.$$

Adding up, we obtain estimate a).

Q.E.D.

The hypotheses of Th. 7 are sufficiently general to deal with most applications known so far.It may be noted however that, modifying the constants suitably, the conclusions of the theorem remain valid if we assume merely that $\Omega'' \in L^1(\Sigma)$ and that $\Omega' \in L \log^+ L(\Sigma)$ and has mean value zero on Σ.

The previous result can be restated as follows. Let $K(x)$ be a singular kernel in E^n, with mean value zero on the unit sphere Σ and and with characteristic Ω in $L^q(\Sigma)$ for some $q > 1$. Then, if $f \in L^p(E^n)$, $1 < p < \infty$, the principal value convolution

$$\tilde{f} = Kf = \text{p.v.}(K*f) = \lim_{\varepsilon \to 0} (K_\varepsilon * f) = \lim_{\varepsilon \to 0} \tilde{f}_\varepsilon$$

exists as a limit in L^p norm and defines a bounded linear operator in $L^p(E^n)$.

We conclude this chapter with a brief discussion of some additional results of Calderòn and Zygmund which are also of independent interest. The numbers in [ ] refer to the bibliography at the end of these Notes.

THEOREM 8: Under the same hypotheses of Th. 7, $\tilde{f}(x) = \lim_{\varepsilon \to 0} \tilde{f}_\varepsilon(x)$ exists pointwise almost everywhere.

Proof: Consider the function $\tilde{f}_*$ defined by

$$\tilde{f}_*(x) = \sup_{\varepsilon > 0} |\tilde{f}_\varepsilon(x)|.$$

CLAIM:

(i) $\|\tilde{f}_*\|_p \leqslant C_{p,q}\|f\|_p$.

It can be shown (see [4]) that if $\varepsilon = \varepsilon(x) > 0$ is a positive measurable function, then

(1) $\|\tilde{f}_{\varepsilon(x)}\|_p \leqslant A_{p,q}\|f\|_p$

where the constant $A_{p,q}$ depends now also on Ω. Fix a ball B around the origin. Given any $\delta > 0$ there exists a function $\varepsilon(x) > 0$ such that

$$\big|\, |\tilde{f}_{\varepsilon(x)}(x)| - \tilde{f}_*(x)\big| \leqslant \delta \qquad \text{for almost all } x \in B.$$

Hence

(2) $$\Big\{\int_B \big|\, |\tilde{f}_{\varepsilon(x)}(x)| - \tilde{f}_*(x)\big|^p \, dx\Big\}^{1/p} \leqslant \delta |B|^{1/p}.$$

By (1) and (2) and Minkowski's inequality, it follows that

$$\Big\{\int_B \tilde{f}_*(x)^p \, dx\Big\}^{1/p} \leqslant \delta |B|^{1/p} + A_{p,q}\|f\|_p.$$

If we first let $\delta \to 0$ and then let B expand to E^n, we obtain (i).

Now we consider the difference

$$\Delta(x,f) = \lim_{\varepsilon\to 0} \sup \tilde{f}_\varepsilon(x) - \lim_{\varepsilon\to 0} \inf \tilde{f}_\varepsilon(x)$$

which is defined almost everywhere; in fact $|\Delta(x,f)| \leq 2\tilde{f}_*(x)$. Clearly $\tilde{f}(x) = \lim_{\varepsilon\to 0} \tilde{f}_\varepsilon(x)$ exists if and only if $\Delta(x,f) = 0$. Thus it remains to prove that $\Delta(x,f)$ vanishes a.e. .

For any $\eta > 0$ there exists a function $g \in C^1_o$ such that $\|f-g\|_p<\eta$. Moreover, by Th. 1 of §1, $\tilde{g}(x) = \lim_{\varepsilon\to 0} \tilde{g}_\varepsilon(x)$ exists almost everywhere; hence, a.e. $\Delta(x,g) = 0$ and so $\Delta(x,f) = \Delta(x,f-g)$. Using (i), with f replaced by f-g, we obtain

$$\|\Delta(x,f)\|_p = \|\Delta(x,f-g)\|_p \leq 2\|(\widetilde{f-g})_*\|_p \leq 2C_{p,q}\|f-g\|_p < 2C_{p,q}\eta$$

and we can conclude that $\|\Delta(x,f)\|_p = 0$. Therefore $\Delta(x,f) = 0$ a.e.

Q.E.D.

If $p = 1$ there are analogues of Theorems 5 and 7 of Chapter III which we summarize in the following statement.

THEOREM 9: Suppose that the kernel $K(x)$ satisfies, in addition to the hypotheses of Th. 7, the following condition:

$$\int_{|x|>2|y|} |K(x-y)-K(x)|\,dx \leq C$$

for some constant $C > 0$ independent of y. If $f \in L^1(E^n)$, then

(a) $\tilde{f}(x)$ exists pointwise almost everywhere.

(b) The operator $Kf = \tilde{f}$ is of weak-type (1,1).

(c) $\tilde{f} \in L^{1-\alpha}(B)$ for any $0 < \alpha < 1$ and any bounded subset B of E^n.

For the proofs of these results see [3] and also §2 of [8].

The composition of two singular integral operators* of the same form since the mean value zero condition may no longer hold. The situation, however, can be easily remedied by considering operators of the form

* with kernels satisfying the hypotheses of Th. 7 is not a singular integral operator

$$Hf = \alpha f + Kf$$

where α is a complex number and K is as before. Algebras of singular integral operators of this kind have been studied in [5].

Finally, we mention that much of the recent research on singular integral operators has focused on operators with variable kernels:

$$Kf(x) = \lim_{\varepsilon \to 0} \int_{|x-y|>\varepsilon} K(x,x-y)f(y)\, dy$$

where $K(x,z) = \Omega(x,z')|z|^{-n}$ and, for each x, Ω has mean value zero on $\Sigma = \{z: |z| = 1\}$. Under mere integrability conditions on Ω, one has the following result whose proof (somewhat complicated but based on ideas we have already employed) can be found in [4].

THEOREM 10: Let K(x,z) be as above and suppose that, for $1 < q < \infty$,

$$\int_{\Sigma} |\Omega(x,z')|^q\, dz'$$

is finite and bounded as a function of x. If $f \in L^p(E^n)$, with $\infty > p \geq q/(q-1)$, then Kf exists pointwise almost everywhere and as a limit in L^p norm, and defines a bounded linear operator in $L^p(E^n)$.

However, for the purpose of applications to partial differential equations, it suffices to consider kernels K(x,z) which (besides the usual homogeneity and mean value zero condition) are infinitely differentiable with respect to z, in $|z| > 0$, somewhat differentiable in x, and such that all the derivates in question are bounded on $E^n \times \Sigma$. The study of the corresponding operators is carried out in [6], and summarized in §6 of [8]. This investigation has proved very fruitful and has given rise to a variety of new and exciting developments in Analysis as well as in Topology.

EXERCISES

1. Prove that, for any bounded set $E \subset E^n$, $L^1(E) \supset L\log^+L(E) \supset L^p(E)$, $1 < p \leq \infty$.

2. A kernel $K(x)$, satisfying the hypotheses of Th. 7, is said to be a smooth singular kernel if $K(x) \in C^\infty$, in $|x| > 0$. A function $h(x) \in C^\infty$ in $|x| > 0$, homogeneous of degree zero and with mean value zero on Σ, is called a smooth symbol. As proved in [5], the Fourier transform gives a 1-1 correspondence between smooth singular kernels and smooth symbols. Assuming this, show that the class $\mathcal{A}$ of operators of the form

$$Hf = \alpha f + Kf$$

is a commutative algebra.

3. Prove that an operator $H \in \mathcal{A}$ has an inverse in $\mathcal{A}$ if and only if its (smooth) symbol, $h(x) = \alpha + \hat{K}(x)$, never vanishes.

BIBLIOGRAPHY

[1] M. Riesz, "Sur les fonctions conjuguées", Math. Zeit., 27, (1927), 218-244.

[2] S.G. Mihlin, "Singular Integrals", Amer. Math. Soc. Translation No. 24, (1950), 12-107.

[3] A.P. Calderòn and A. Zygmund, "On the existence of certain singular integrals", Acta Math., 88, (1952), 85-139.

[4] ______________________, "On Singular Integrals", Amer. J. Math., 78, (1956), 289-309.

[5] ______________________, "Algebras of certain singular operators", Amer. J. Math., 78, (1956), 310-320.

[6] ______________________, "Singular integral operators and differential equations", Amer. J. Math., 79, (1957), 901-921.

[7] A. Zygmund, "Intégrales Singulières", Université de Paris (Orsay), (1965).

[8] A. P. Calderòn , "Singular Integrals", Bull. Amer. Math. Soc., 72, (1966), 427-465

PART II

SINGULAR INTEGRAL OPERATORS AND DISTRIBUTIONS

To Alberto Calderón
who taught me all this
and much more

PREFACE

The purpose of these lectures, which are a continuation of "Singular Integrals: An Introduction" (University of Maryland, Lecture Notes No. 3, 1967), is twofold: first, to extend the basic continuity properties of singular integral operators to a general class of Sobolev spaces; secondly, to discuss a larger class of singular integral operators (not necessarily of convolution type) which plays a basic role in the theory of singular integrals.

In preparing these lectures for a course I gave last term at the University of Maryland, I relied heavily on the excellent notes from a seminar held at the University of Buenos Aires by Professor Alberto P. Calderón ([1]), as well as on a classic paper by Calderón and A. Zygmund ([2]). (Numbers in [ ] refer to the bibliography at the end of these notes.) Moreover, in order to make this material more easily accessible to students, I devoted the first chapter to an elementary survey of some basic notions from the theory of distributions.

I wish to thank Mrs. Judith Boyce and Miss Susan Gates for their accurate typing and Mr. Robert J. Hill who directed the task of publishing this manuscript.

U. Neri

College Park, August 1968

NOTATION

We denote by $\mathbb{R}$ and $\mathbb{C}$ the real and complex numbers respectively. Euclidean n-dimensional space E^n is the set of all elements $x = (x_1, \ldots, x_n)$ in $\mathbb{R}^n$, with the inner product $(x,y) = x_1y_1 + \ldots + x_ny_n$ and the corresponding norm $|x| = (x,x)^{1/2}$. On E^n, we let $dx = dx_1 \ldots dx_n$ denote the Lebesgue measure. All sets and functions considered will be Lebesgue measurable and

$$\int f(x)\,dx = \int \ldots \int f(x_1, \ldots, x_n)\,dx_1 \ldots dx_n$$

will denote the (Lebesgue) integral of the function f over the entire space E^n. Unless otherwise stated, all functions are complex valued.

It is often convenient to express a multiple integral in "polar coordinates", i.e. letting $r = |x|$, $\Sigma = \{x: |x| = 1\}$ the unit sphere, we can write the element of volume dx in the form $dx = r^{n-1}\,dr\,d\sigma$, where $d\sigma$ is the surface measure on Σ induced by dx. Then, if $f(x) \geq 0$ is an integrable function, say, we have by Fubini's theorem that

$$\int f(x)\,dx = \int_0^\infty \left\{ \int_\Sigma f(x)\,d\sigma \right\} r^{n-1}\,dr = \int_\Sigma \left\{ \int_0^\infty f(x)\,r^{n-1}\,dr \right\} d\sigma.$$

If $x \neq 0 = (0, \ldots, 0)$, then $x = |x| \frac{x}{|x|} = rx'$, where $x' = x/|x|$ is the projection of x on Σ. Since we often denote by x' the variable on Σ, it is convenient in those instances to write dx' in place of $d\sigma$.

By a multi-index α we mean an n-tuple $\alpha = (\alpha_1, \ldots, \alpha_n)$ of integers $\alpha_i \geq 0$, and we write $|\alpha| = \alpha_1 + \ldots + \alpha_n$, $\alpha! = \alpha_1! \ldots \alpha_n!$, $\alpha + \beta = (\alpha_1 + \beta_1, \ldots, \alpha_n + \beta_n)$. Furthermore, we let $x^\alpha = x_1^{\alpha_1} x_2^{\alpha_2} \ldots x_n^{\alpha_n}$ and, similarly, $\left(\frac{\partial}{\partial x}\right)^\alpha = \frac{\partial^{\alpha_1}}{\partial x_1^{\alpha_1}} \frac{\partial^{\alpha_2}}{\partial x_2^{\alpha_2}} \ldots \frac{\partial^{\alpha_n}}{\partial x_n^{\alpha_n}}$. Often, it is convenient to use the differentiations $D = (D_1, \ldots, D_n)$ where $D_j = \frac{1}{2\pi i} \frac{\partial}{\partial x_j}$ and $D^\alpha = (2\pi i)^{-|\alpha|} \left(\frac{\partial}{\partial x}\right)^\alpha$.

We shall always take the Fourier transform of a function f in the form

$$(Ff)(\xi) = \hat{f}(\xi) = \int e^{-2\pi i(x,\xi)} f(x)\, dx$$

where x and ξ are dual variables according to the bilinear pairing $(x,\xi) = x_1\xi_1 + \ldots + x_n\xi_n$.

CHAPTER I

DISTRIBUTIONS AND FOURIER TRANSFORMS

§1. DISTRIBUTIONS

Let us denote by $C_o^\infty = C_o^\infty(E^n)$ the class of smooth functions $u:E^n \to \mathbb{C}$ with compact support. Clearly, C_o^∞ is a (complex) vector space and we may consider the topology induced on it by the following convergence criterion: we say that u_n converges-S to u (in symbols, $u_n \rightarrowtail u$) if the u_n have support in a fixed compact set K and u_n converge to u uniformly together with all their derivatives. More precisely, for every multi-index α,

$$\left(\frac{\partial}{\partial x}\right)^\alpha u_n \longrightarrow \left(\frac{\partial}{\partial x}\right)^\alpha u$$

uniformly on K. Equipped with this topology, C_o^∞ becomes a topological vector space (t.v.s.) which will be denoted by $\mathcal{D}$.

DEFINITION 1.1.0. The t.v.s. $\mathcal{D}^*$ of all continuous conjugate linear functionals on $\mathcal{D}$ is called the space of (Schwartz) distributions on E^n.

If $\ell \in \mathcal{D}^*$ and $u \in \mathcal{D}$ we write $\langle \ell, u \rangle$ for the complex number $\ell(u)$ expressing the value of ℓ on the element u. As we shall see presently, $\mathcal{D}^*$ is very rich in elements; in fact every locally integrable function can be regarded as an element of $\mathcal{D}^*$. Furthermore, we can define on $\mathcal{D}^*$ a large number of algebraic and analytic operations.

The basic properties of $\mathcal{D}^*$ can be summarized as follows:

P.1.1.1. $\mathcal{D}^*$ is a vector space containing $L^1_{loc}(E^n)$.

In fact if f is locally integrable, then for all u in $\mathcal{D}$ the functional $\langle f,u\rangle = \int f\bar{u}\,dx$ is defined. In addition, it is easy to see that $u_n \rightarrowtail 0$ implies that $\langle f,u_n\rangle \longrightarrow 0$, so $f \in \mathcal{D}^*$.

P.1.1.2. $\mathcal{D}^*$ is a module over $C^\infty = C^\infty(E^n)$, where we define $\langle uf,v\rangle = \langle f,\bar{u}v\rangle$ for all $f \in \mathcal{D}^*$, $u \in C^\infty$ and $v \in \mathcal{D}$.

Noticing that if $f \in L^1_{loc}$ and $u \in \mathcal{D}$ then

$$\langle \bar{f},u\rangle = \int \bar{f}\,\bar{u}\,dx = \overline{\int f\,u\,dx} = \overline{\langle f,\bar{u}\rangle},$$

we define, for all $g \in \mathcal{D}^*$ and $u \in \mathcal{D}$, $\langle \bar{g},u\rangle = \overline{\langle g,\bar{u}\rangle}$.

P.1.1.3. $f \in \mathcal{D}^*$ implies that $\bar{f} \in \mathcal{D}^*$.

Similarly if f is locally integrable and we define $\check{f}(x) = \overline{f(-x)}$, then for every u in $\mathcal{D}$ we have

$$\langle \check{f},u\rangle = \int \check{f}\,\bar{u}\,dx = \int \overline{f(-x)}\;\overline{u(x)}\,dx = \int \overline{f(y)}\;\overline{u(-y)}\,dy =$$

$$= \overline{\int f(y)\,\overline{\overline{u(-y)}}\,dy} = \overline{\int f(y)\,\overline{\check{u}(y)}\,dy} = \overline{\langle f,\check{u}\rangle}.$$

Thus given any g in $\mathcal{D}^*$ we define $\check{g}$ by setting $\langle \check{g},u\rangle =$ $= \overline{\langle g,\check{u}\rangle}$ for all u in $\mathcal{D}$, and we obtain

P.1.1.4. $f \in \mathcal{D}^*$ implies that $\check{f} \in \mathcal{D}^*$.

Given an f in L^1_{loc}, the translate $(\tau_a f)(x) = f(x - a)$ is again locally integrable. Furthermore, for all u in $\mathcal{D}$,

$$\langle \tau_a f,u\rangle = \int f(x - a)\,\overline{u(x)}\,dx = \int f(y)\,\overline{u(y + a)}\,dy = \langle f,\tau_{-a}u\rangle$$

so we define the translate $\tau_a g$ of a distribution $g \in \mathcal{D}^*$ by the formula $\langle \tau_a g, u \rangle = \langle g, \tau_{-a} u \rangle$ for all $u \in \mathcal{D}$, and we have

P.1.1.5. $f \in \mathcal{D}^*$ implies that $\tau_a f \in \mathcal{D}^*$.

Letting $D_j = \frac{1}{2\pi i} \frac{\partial}{\partial x_j}$ we note that if $f \in C^1(E^n)$ and $u \in \mathcal{D}$ then

$$\langle D_j f, u \rangle = \frac{-i}{2\pi} \int \frac{\partial f}{\partial x_j} \bar{u} \, dx = \text{(integrating by parts)}$$

$$= \frac{i}{2\pi} \int f \frac{\partial \bar{u}}{\partial x_j} dx = \int f \, \overline{D_j u} \, dx = \langle f, D_j u \rangle .$$

Therefore, given g in $\mathcal{D}^*$ we define $D_j g$ by the formula $\langle D_j g, u \rangle = \langle g, D_j u \rangle$ for all u in $\mathcal{D}$ and we obtain

P.1.1.6. $f \in \mathcal{D}^*$ implies that $D_j f \in \mathcal{D}^*$.

N.B.: An equivalent definition of the distribution derivatives $D_j f$ can be given using P.1.1.5. and the fact that $\mathcal{D}^*$ is a t.v.s. with the topology of weak convergence; i.e., given f and f_n in $\mathcal{D}^*$ we say that $f_n \longrightarrow f$ (weakly) in $\mathcal{D}^*$ if, for all u in $\mathcal{D}$, $\langle f_n, u \rangle \longrightarrow \langle f, u \rangle$ as $n \longrightarrow \infty$.

P.1.1.7. The support of a distribution.

Given f in $\mathcal{D}^*$, we say that $f = 0$ on an open set $U \subset E^n$ if $\langle f, u \rangle = 0$ for all $u \in \mathcal{D}$ with support contained in U. We say that $f = g$ on U if $(f - g) = 0$ on U. Now the support, supp f, of a distribution f can be defined as follows:

$x \in E^n$ is a <u>non-essential point</u> of f if there exists an open neighborhood V_x such that $f = 0$ on V_x. Otherwise x is said to be an <u>essential point</u> of f. We now set supp f = the set of all essential points of f. It is easy to see that supp f is a closed set. We shall denote by $\mathcal{D}_o^*$ the vector subspace of distributions with compact support.

We have next two representation theorems for distributions.

THEOREM 1.1.8. Let f be in $\mathcal{D}^*$. Then there exists a sequence $\{f_k\}$ in $\mathcal{D}_o^*$ with the following properties:

(i) for any bounded subset A of E^n, $(\text{supp } f_k) \cap A = \emptyset$ except for finitely many k.

(ii) for all u in $\mathcal{D}$, $\langle f,u \rangle = \Sigma_k \langle f_k,u \rangle$.

N.B. The sum in (ii) is finite, for each u, because of (i).

THEOREM 1.1.9. If f is in $\mathcal{D}_o^*$ then there exist a function ψ in $\mathcal{D}$ and a continuous function g in $C^o = C^o(E^n)$ such that, for some multi-index α, $f = \psi D^\alpha g$ where $D^\alpha g$ are distribution derivatives; i.e., since $C^o \subset L^1_{loc} \subset \mathcal{D}^*$, $D^\alpha g$ are the derivatives of g as an element of $\mathcal{D}^*$.

Proof of 1.1.8. Suppose that we have a smooth partition of unity $\{\phi_k\}$ in $\mathcal{D}$ such that for any bounded set A in E^n $(\text{supp } \phi_K) \cap A = \emptyset$ except for finitely many k and, of course, $\Sigma_k \phi_k(x) = 1$ for all x in E^n. Then the distributions f_k, given by $f_k = \phi_k f$, have support contained in $\text{supp } \phi_k$ and hence satisfy (i). Moreover, for all u in $\mathcal{D}$, we obtain, using P.1.1.2.,

$$\langle f,u \rangle = \Sigma_k \langle f,\bar{\phi}_k u \rangle = \Sigma_k \langle \phi_k f,u \rangle = \Sigma_k \langle f_k,u \rangle$$

where the summation has only a finite number of non-zero terms.

To complete the proof it remains to construct the desired partition of unity $\{\phi_k\}$. Consider a function $\eta(t) \geq 0$ in

$C_o^\infty(E^1)$ such that $\eta(t) > 0$ if $|t| < 1$, $\eta(t) = 0$ if $|t| \geq 1$. Let $\{x^{(k)}\}$ be an enumeration of the lattice points in E^n, i.e. the points of E^n with integral coordinates, and denote by x_j the jth coordinate of $x \in E^n$. Now, the functions

$$\psi_k(x) = \prod_{j=1}^n \eta(x_j - x_j^{(k)})$$

are ≥ 0 and belong to $C_o^\infty(E^n)$. Moreover, their sum $\psi(x) = \Sigma_k \psi_k(x)$ belongs to $C^\infty(E^n)$ and $\psi(x) > 0$ everywhere. Finally we let $\phi_k = \psi_k/\psi$ and verify that these functions have all the desired properties. Q.E.D.

Proof of 1.1.9. We must show here that every distribution f with compact support is essentially a distribution derivative of some continuous function.

Given f in $\mathcal{D}_o^*$ let us choose an $M > 0$ large enough so that supp f is contained in the ball $K = \{x: |x| \leq M\}$. Introduce the notation $\mathcal{D}(K) = \{u \in \mathcal{D} : (\text{supp } u) \subset K\}$. For every integer $m \geq 0$ and all u in $\mathcal{D}(K)$, define the quantity $p_m(u)$ by

$$(1) \qquad p_m(u) = \sup_{x \in K} \left\{ \sup_{|\alpha| \leq m} |D^\alpha u| \right\}$$

and note that $\{p_m\}$ is a sequence of semi-norms on $\mathcal{D}(K)$. Now we want to show that f is continuous with respect to a semi-norm p_m, for some $m \geq 0$. Since f is a linear functional, we must prove that for some integer $m \geq 0$ there exists a constant $C > 0$ such that, for all u in $\mathcal{D}(K)$,

$$(2) \qquad |\langle f,u\rangle| \le Cp_m(u).$$

Formula (2) indeed holds since otherwise we would have a sequence $\{u_m\}$ in $\mathcal{D}(K)$ with $p_m(u_m) \le 1/m$ whereas $|\langle f,u_m\rangle| \ge 1$. But then $u_m \longrightarrow 0$ in the topology induced by $\{p_m\}$ and so, by (1), also in the topology of $\mathcal{D}$ and hence, since $f \in \mathcal{D}^*$, $\langle f,u_m\rangle \longrightarrow 0$ which is a contradiction.

From now on, m is a fixed non-negative integer for which (2) holds for all u in $\mathcal{D}(K)$.

CLAIM: If $\alpha = (m + n, \ldots, m + n)$ is an n-tuple, then there exists a new constant $c > 0$ such that, for all u in $\mathcal{D}(K)$,

$$(3) \qquad p_m(u) \le c \int |D^\alpha u| \, dx.$$

This formula follows easily from the expression

$$u(x) = \int_{-\infty}^{x_1} \cdots \int_{-\infty}^{x_n} D^{(1,\ldots,1)} u(y) \, dy_n \ldots dy_1$$

if we bound the derivatives $D^\alpha u$ using repeated integrations by parts. The details are left as an exercise.

Combining (2) and (3) we obtain a new constant $c_1 > 0$ such that for all u in $\mathcal{D}(K)$

$$(4) \qquad |\langle f,u\rangle| \le c_1 \int |D^\alpha u| \, dx, \qquad \alpha = (m + n, \ldots, m + n).$$

Formula (4) shows that f defines a bounded linear functional on the subspace of the Banach space $L^1(K)$ consisting of all u in $L^1(K)$ such that $D^\alpha u$ also belongs to $L^1(K)$. Therefore the Hahn-Banach theorem implies that f extends to a bounded linear functional on all of $L^1(K)$ and the extension has the same norm. But, since the conjugate space of $L^1(K)$ is precisely $L^\infty(K)$, it follows that there exists a function h in $L^\infty(K)$ such that, for all u in $\mathcal{D}(K)$,

$$\langle h, D^\alpha u \rangle = \langle f, u \rangle.$$

Since $L^\infty \subset L^1_{loc} \subset \mathcal{D}^*$, viewing h as a distribution we obtain, using P.1.1.6., that for all u in $\mathcal{D}(K)$

$$\langle D^\alpha h, u \rangle = \langle f, u \rangle. \tag{5}$$

Since $(\text{supp } f) \subset K$ and $h \in L^\infty(K)$, we can view h as an element of $L^\infty(E^n)$ by setting $h(x) = 0$ if x is not in K. Now, to obtain the desired function g, we integrate h. Specifically, we consider the (multiple) integral

$$g(x) = (2\pi i)^n \int_{-\infty}^{x} h(y)\, dy$$

which is well-defined because h is now an integrable function. Moreover $h(x) = D_1 \cdot\cdot\cdot D_n g(x)$ almost everywhere and hence as distribution, so from (5) we deduce that

(6) $$\langle D^{\alpha}D_1 \ldots D_n g,u\rangle = \langle D^{\alpha}h,u\rangle = \langle f,u\rangle$$

for all u in $\mathcal{D}(K)$.

Finally we verify that for some function ψ in $\mathcal{D}$ we have $f = \psi D^{\beta}g$ as elements of $\mathcal{D}^*$. To do this we may assume from the beginning that $(\operatorname{supp} f) \subset \{x: |x| \leq M - 1\} = K_1$ whereas $K = \{x: |x| \leq M\}$. We choose next a function ψ in $\mathcal{D}$ such that $\psi(x) = 1$ if $x \in K_1$ and $\psi(x) = 0$ if $x \notin K$. Since $u \in \mathcal{D}$ implies that $\bar{\psi}u \in \mathcal{D}(K)$, using (6) we obtain on the one hand that

$$\langle f,\bar{\psi}u\rangle = \langle D^{\beta}g,\bar{\psi}u\rangle = \langle \psi D^{\beta}g,u\rangle$$

and, on the other hand,

$$\langle f,\bar{\psi}u\rangle = \langle \psi f,u\rangle = \langle f,u\rangle$$

for all u in $\mathcal{D}$. Therefore $f = \psi D^{\beta}g$. Q.E.D.

EXAMPLE 1.1.10. The Dirac measure (δ-function).

On E^1, let us consider the Heaviside function $H(t) = 0$ if $t \leq 0$ and $H(t) = 1$ if $t > 0$. Clearly $H(t)$ is locally integrable, so it can be viewed as a distribution and in this sense we consider the derivative $DH(t) = \frac{1}{2\pi i}\frac{d}{dt}H(t)$. For all u in $\mathcal{D} = \mathcal{D}(E^1)$, we have, by P.1.1.6. and integration, that

$$\langle DH,u\rangle = \langle H,Du\rangle = \frac{i}{2\pi}\int_0^{\infty} \overline{u'(t)}\,dt = \frac{1}{2\pi i}\,\overline{u(0)},$$

so if we define the distribution δ, on E^n, by the formula

$$\langle \delta, u \rangle = \overline{u(0)} \quad , \qquad u \in \mathcal{D} = \mathcal{D}(E^n), \quad n \geq 1,$$

we obtain that $DH = \frac{1}{2\pi i}\,\delta$, i.e. $\delta = \frac{d}{dt} H$ on E^1.

P.1.1.11. Convolution of distributions with functions in $\mathcal{D}$.

Let f be locally integrable and u in $\mathcal{D}$. Then the convolution $f * u$ exists and furthermore

$$(f * u)(x) = \int f(x - y)\, u(y)\, dy = \int f(y)\, u(x - y)\, dy =$$
$$= \int f(y)\, \overline{\check{\bar{u}}(y - x)}\, dy = \langle f, \tau_x \check{u} \rangle$$

which is a smooth function of x as can be seen by differentiating under the integral sign.

Therefore, given now f in $\mathcal{D}^*$ and u in $\mathcal{D}$ we define the convolution $f * u$ to be the function given by

(a) $$(f * u)(x) = \langle f, \tau_x \check{u} \rangle .$$

Alternatively, viewing $f \in L^1_{loc}$ as a distribution we have for all u and v in $\mathcal{D}$

$$\langle f * u, v \rangle = \int (f * u)\,(x)\, \overline{v(x)}\, dx =$$
$$= \int \left\{ \int f(x - y)\, u(y)\, dy \right\} \overline{v(x)}\, dx =$$
$$= \int \left\{ \int f(y)\, u(x - y)\, dy \right\} \overline{v(x)}\, dx = \text{(Fubini's Theorem)}$$

$$= \int f(y) \left\{ \int u(x - y) \overline{v(x)} \, dx \right\} dy =$$

$$= \int f(y) \left\{ \overline{\int \check{u}(y - x) v(x) \, dx} \right\} dy =$$

$$= \langle f, \check{u} * v \rangle.$$

In other words we can also define, if $f \in \mathcal{D}^*$ and $u \in \mathcal{D}$, the distribution $f * u$ by the formula

(b) $$\langle f * u, v \rangle = \langle f, \check{u} * v \rangle \quad , \qquad v \in \mathcal{D}.$$

As shown in various texts (e.g. in [7]) definitions (a) and (b) are equivalent.

P.1.1.12. Convolution of distributions with distributions in $\mathcal{D}^*_o$.

We note first of all that distributions with compact support are linear functionals on $C^\infty = C^\infty(E^n)$ which are continuous with respect to the topology put on C^∞_o earlier. In fact, if $f \in \mathcal{D}^*_o$ and ϕ in any real valued function in $\mathcal{D}$ such that $\phi(x) = 1$ on a neighborhood of supp f, then $\phi f = f$ and therefore for all $u \in C^\infty$

$$\langle f, u \rangle = \langle \phi f, u \rangle = \langle f, \phi u \rangle$$

where now ϕu is in $\mathcal{D}$.

N.B. The class C^∞ with uniform convergence of all derivatives on compacts is often denoted by $\mathcal{E}$. Consequently $\mathcal{D}^*_o$ is often denoted by $\mathcal{E}^*$.

Suppose now that f is in $\mathcal{D}_o^*$ and g belongs to $\mathcal{D}^*$. For all $u \in \mathcal{D}$, we have two possible definitions of convolution:

(a) $$\langle g * f, u \rangle = \langle g, \check{f} * u \rangle$$

(b) $$\langle f * g, u \rangle = \langle f, \check{g} * u \rangle \ .$$

Indeed both (a) and (b) make sense. In (a) we observe that $(\text{supp}\ \check{f} * u) \subset \{x + y : x \in \text{supp}\ f \text{ and } y \in \text{supp}\ u\}$ so that by the previous property $\check{f} * u$ belongs to $\mathcal{D}$ and hence $\langle g, \check{f} * u \rangle$ is well defined. In (b) we note that $\check{g} * u$ is in C^∞ and so, by our initial remark, $\langle f, \check{g} * u \rangle$ is well defined.

Moreover it can be shown that (a) and (b) are equivalent (e.g. [7]); in other words convolution, as for ordinary scalar valued functions, is cummutative. In other words, for all $f \in \mathcal{D}_o^*$ and g in $\mathcal{D}^*$, $f * g = g * f$. Furthermore, we may consider the distribution derivatives $D^\alpha(f * g)$ of the distribution $f * g$. For all $u \in \mathcal{D}$ we have

$$\langle D^\alpha(f * g), u \rangle = \langle f * g, D^\alpha u \rangle = \langle f, \check{g} * D^\alpha u \rangle =$$

$$= \langle f, (D^\alpha \check{g}) * u \rangle = \langle f, D^\alpha(\check{g} * u) \rangle .$$

Now, on the one hand,

$$\langle f, D^\alpha(\check{g} * u) \rangle = \langle D^\alpha f, \check{g} * u \rangle = \langle (D^\alpha f) * g, u \rangle$$

while, on the other hand,

$$\langle f, D^{\alpha}(\check{g} * u)\rangle = \langle f, (D^{\alpha}\check{g}) * u\rangle = \langle f * (D^{\alpha}g), u\rangle .$$

Therefore,

$$D^{\alpha}(f * g) = (D^{\alpha}f) * g = f * (D^{\alpha}g).$$

REMARK 1.1.13. The Dirac measure δ belongs to $\mathcal{D}_o^*$ since it has "point support" at the origin of E^n. So, for any $f \in \mathcal{D}^*$, $\delta * f$ is again a distribution. Moreover,

$$(*) \qquad \delta * f = f.$$

Proof of (*). For all u in $\mathcal{D}$, we have that

$$\langle \delta * f, u\rangle = \langle \delta, \check{f} * u\rangle = \overline{(\check{f} * u)(0)} = \quad \text{[by 1.1.11(a)}$$

with $x = 0$] $\qquad = \overline{\langle \check{f}, \check{u}\rangle} = \langle f, u\rangle .$

Similarly, one has that $f * \delta = f$. Therefore δ acts as a (2-sided) identity under the "convolution product" of distributions.

We may recall here that if f is a bounded continuous function and $K(x) = K(|x|) \geq 0$ is a "bell-shaped" function on E^n with support in $|x| \leq 1$ and such that $\int K(x)\, dx = 1$, then letting

$$K_\varepsilon(x) = \varepsilon^{-n} K\left(\frac{x}{\varepsilon}\right)$$

we have that, at each point x,

$$(K_\varepsilon * f)(x) = (f * K_\varepsilon)(x) \longrightarrow f(x)$$

as $\varepsilon \to 0$. Hence the functions K_ε are sometimes called "approximate identities". Indeed, in the topology of $\mathcal{D}^*$,

$$\delta = \lim_{\varepsilon \to 0} K_\varepsilon(x).$$

P.1.1.14. $\mathcal{D}$ is dense in $\mathcal{D}^*$, with respect to the (weak) topology of $\mathcal{D}^*$. In other words, every f in $\mathcal{D}^*$ is the weak limit of a sequence f_m in $\mathcal{D}$.

Proof. Fix f in $\mathcal{D}^*$ and choose a sequence ψ_m in $\mathcal{D}$ such that $\psi_m(x) \longrightarrow 1$ pointwise as $m \longrightarrow \infty$, e.g. $\psi_m(x) = 1$ if $|x| \le m$. Then, for any $u \in \mathcal{D}$ and all m so large that (supp $u) \subset \{x: |x| \le m\}$, we have that

$$\langle \psi_m f, u \rangle = \langle f, \bar{\psi}_m u \rangle = \langle f, u \rangle.$$

Next we pick $\phi \in \mathcal{D}$ such that $\phi(x) = 1$ near the origin, $\phi(x) = 0$ if $|x| \ge 1$, and $\int \phi(x)\, dx = 1$. When we form the approximate identities $\phi_\varepsilon(x) = \varepsilon^{-n} \phi(x/\varepsilon)$ we see that ϕ_ε belong to $\mathcal{D}$ and $\phi_\varepsilon \longrightarrow \delta$, weakly, as $\varepsilon \longrightarrow 0$. Thus, taking $\varepsilon = 1/m$ and considering the convolutions

$$f_m = (\psi_m f) * \phi_{1/m}$$

we note that $f_m \in \mathcal{D}$ since $\psi_m f \in \mathcal{D}_o^*$ so, using 1.1.11, convolution with $\phi_{1/m}$ in $\mathcal{D}$ yields a function in $\mathcal{D}$. Moreover, it is easy to verify that for all $u \in \mathcal{D}$

$$\langle f_m, u \rangle \longrightarrow \langle f, u \rangle \quad \text{as} \quad m \longrightarrow \infty.$$

Q.E.D.

Finally, we mention an important characterization of distributions with (compact) support at the origin:

P.1.1.15. If $f \in \mathcal{D}^*$ and $(\text{supp } f) = 0 \in E^n$, then f is a finite linear combination of derivatives of the Dirac measure δ.

The proof of this result may be found for example in [5].

§2. TEMPERATE DISTRIBUTIONS AND FOURIER TRANSFORM

For many purposes, including our own, the class of distributions $\mathcal{D}^*$ is too big whereas the class $\mathcal{D}^*_o$ of distributions with compact support is too small. So we must introduce the class $\mathcal{S}^*$, $\mathcal{D}^*_o \subset \mathcal{S}^* \subset \mathcal{D}^*$, of so-called temperate distributions.

Let us denote by $\mathcal{S}$ the vector space of C^∞ functions which, together with all their derivatives, approach zero, as $|x| \longrightarrow \infty$, faster than any negative power of $|x|$. For example, the function $\exp(-|x|^2)$ belongs to $\mathcal{S}$. More precisely $u \in \mathcal{S}$ if for all multi-indices α and β there exist positive constants $C_{\alpha\beta}$ such that

$$(1) \qquad |x^\alpha D^\beta u(x)| \le C_{\alpha\beta}.$$

Equivalently, for all $|\beta| \geq 0$ and every number $s \geq 0$,

$$(1') \qquad |D^\beta u(x)| \le C_{s\beta}(1 + |x|^2)^{-s}.$$

Functions in $\mathcal{S}$ are said to be smooth, rapidly decreasing functions.

Convergence in $\mathcal{S}$ is defined as follows. A sequence $\{u_K\}$ in $\mathcal{S}$ is said to converge to u if, on every compact $K \subset E^n$ and for all $|\alpha| \geq 0$, $D^\alpha u_k \longrightarrow D^\alpha u$ uniformly on K and moreover the constants $C_{\alpha\beta}$ for which

$$(2) \qquad |x^\beta D^\alpha u_k(x)| \le C_{\alpha\beta}$$

can be chosen independent of k. Then, letting $k \longrightarrow \infty$ in (2) we see that the limit function u also belongs to $\mathcal{S}$.

Definition 1.2.0. The t.v.s. $\mathcal{S}^*$ of all conjugate linear functionals f on $\mathcal{S}$ which are continuous, in the sense that if u_k converges to u in $\mathcal{S}$ then $\langle f,u_k \rangle \longrightarrow \langle f,u \rangle$, is called the space of temperate distributions.

PROPOSITION 1.2.1. The class $\mathcal{S}$ is an algebra of integrable functions which is closed under the operations of differentiation and multiplication by polynomials. If $\mathcal{E}$ denotes the class of C^∞ functions with the uniform convergence of all their derivatives on compact sets, we have that $\mathcal{D} \subset \mathcal{S} \subset \mathcal{E}$, consequently by duality, $\mathcal{E}^* = \mathcal{D}_o^* \subset \mathcal{S}^* \subset \mathcal{D}^*$, where the inclusions are proper, dense and continuous.

Proof. The first assertion is a direct consequence of the conditions (1) and (1'). The function $u(x) = \exp(-|x|^2)$ belongs to $\mathcal{S}$ and $\mathcal{S}^*$ but is not in $\mathcal{D}$ nor in $\mathcal{D}_o^*$. The function $v(x) = \exp(|x|^2)$ belongs to $\mathcal{E}$ and to $\mathcal{D}^*$ but neither to $\mathcal{S}$ nor $\mathcal{S}^*$. So all the inclusions are proper. The continuity of the inclusion maps can be seen directly by noticing that convergence in $\mathcal{D}$ is stronger than convergence in $\mathcal{S}$, and that convergence in $\mathcal{S}$ is stronger than convergence in $\mathcal{E}$.

To verify the density of the inclusions we will show for example that $\mathcal{D}$ is dense in $\mathcal{S}$. Let $\phi_1 \in \mathcal{D}$ be such that $\phi_1(x) = 1$ for $|x| \leq 1$ and $\phi_1(x) = 0$ if $|x| \geq 2$, and consider the sequence $\phi_k(x) = \phi_1(x/k)$. Then, given any u in $\mathcal{S}$, the functions $u_k = u\phi_k$ belong to $\mathcal{D}$ and one easily sees that the

u_k converge to u in $\mathcal{S}$. Q.E.D.

It is a classical result that the Fourier transform

$$(Fu)(\xi) = \hat{u}(\xi) = \int e^{-2\pi i(x,\xi)} u(x)\, dx$$

establishes an isometric isomorphism of the space $L^2 = L^2(E^n)$ with itself. Here x and ξ are dual variables, according to the bilinear pairing $(x,\xi) = x_1\xi_1 + \cdots + x_n\xi_n$. Thus, strictly speaking, we have two copies of E^n, the x-copy and the ξ-copy, and consequently two copies of $L^2(E^n)$.

Our space $\mathcal{S}$, containing $\mathcal{D}$, is evidently dense in L^2. In fact, the Fourier transform, unlike on L^2, can be defined <u>directly</u> and, with $u \in \mathcal{S}$, it is evident that $\hat{u}(\xi)$ exists everywhere in view of the integrability of u. Indeed, $\hat{u}$ is bounded (by the L^1 norm of u), uniformly continuous, and $\hat{u}(\xi) \longrightarrow 0$ as $|\xi| \longrightarrow \infty$. However, before we pass to a more precise description of the role of the Fourier transform on the class $\mathcal{S}$, it is instructive to look at its more special role on the class $\mathcal{D}$.

For simplicity, we shall consider $\mathcal{D} = \mathcal{D}(E^1)$ though all our considerations apply, with the usual modifications, to functions of several real variables. If $u \in \mathcal{D}(E^1)$ and $(\text{supp } u) \subset \{x: |x| \leq a\}$ for instance, then

$$\hat{u}(\xi) = \int_{-a}^{a} e^{-2\pi i x\xi} u(x)\, dx.$$

Furthermore, $\hat{u}$ can be extended to a function of a complex variable $\zeta = \xi + i\eta$ by letting

$$\hat{u}(\zeta) = \int_{-a}^{a} e^{-2\pi i x\zeta} u(x)\, dx = \int_{-a}^{a} e^{-2\pi i x\xi} e^{2\pi x\eta} u(x)\, dx.$$

By differentiating under the integral sign with respect to ζ we see that $\hat{u}(\zeta)$ is holomorphic on the entire ζ-plane. In other words, $\hat{u}(\zeta)$ is an entire analytic function.

The Fourier transform of $Du(x) = \frac{1}{2\pi i} u'(x)$ is seen to be (integrating by parts)

$$\int_{-a}^{a} e^{-2\pi i x\zeta} Du(x)\, dx = \int_{-a}^{a} \zeta u(x)\, e^{-2\pi i x\zeta}\, dx = \zeta\hat{u}(\zeta).$$

Taking powers D^k, $k = 0, 1, 2, \ldots$, of u and repeating the argument, we find that

(1) $$F(D^k u)(\zeta) = \zeta^k \hat{u}(\zeta).$$

Since

$$|F(D^k u)(\zeta)| = \left|\int_{-a}^{a} e^{-2\pi i x\zeta} D^k u(x)\, dx\right| \le C_k e^{2\pi a|\eta|}$$

we conclude that the (complex) Fourier transform $\hat{u}(\zeta)$ of any $u \in \mathcal{D}$, which vanishes for $|x| \ge a$, is an entire analytic function of $\zeta = \xi + i\eta$ which, for every $k = 0, 1, 2, \ldots$, satisfies the growth estimates

(2) $$|\zeta^k \hat{u}(\zeta)| \le C_k e^{2\pi a|\eta|} .$$

This assertion, together with its converse (which is also true), is generally known as the Paley Wiener Theorem It shows, in particular, that the Fourier transforms of functions in $\mathcal{D}$ are a very special class of functions indeed. This class, $F(\mathcal{D})$, is usually denoted by Z.

Let us return to the Fourier transform on the class $\mathcal{S} = \mathcal{S}(E^n)$:

(1) $$\hat{u}(\xi) = \int e^{-2\pi i(x,\xi)}\, u(x)\, dx.$$

<u>THEOREM</u> 1.2.2. The Fourier transformation $u \longrightarrow \hat{u} = Fu$ is a continuous map of $\mathcal{S}$ into itself. Moreover, for any $|\alpha| \ge 0$,

(i) $$D^\alpha \hat{u}(\xi) = F((-x)^\alpha u)(\xi)$$

(ii) $$F(D^\alpha u)(\xi) = \xi^\alpha \hat{u}(\xi).$$

<u>Proof</u>. Since $u \in \mathcal{S}$, we can differentiate (1) under the integral sign (because the integral we obtain is in fact uniformly convergent) and we get

(2) $$D^\alpha \hat{u}(\xi) = \int e^{-2\pi i(x,\xi)}\, (-x)^\alpha\, u(x)\, dx$$

which shows that $\hat{u} \in C^\infty$ and that formula (i) holds.

Multiplying (2) by ξ^β and integrating by parts $|\beta|$-times we obtain

$$(3) \qquad \xi^\beta D^\alpha \hat{u}(\xi) = \int e^{-2\pi i(x,\xi)} D^\beta[(-x)^\alpha u(x)]\, dx$$

so that

$$(4) \qquad |\xi^\beta D^\alpha \hat{u}(\xi)| \le \int |D^\beta[(-x)^\alpha u(x)]|\, dx = C_{\alpha\beta}$$

which shows that $Fu = \hat{u}$ belongs to $\mathcal{S}$. Furthermore, taking $\alpha = (0, \ldots, 0)$ in (3) we see that formula (ii) holds. Finally, replacing u in (4) by a sequence $u_k \longrightarrow 0$ in $\mathcal{S}$, it follows easily that $\hat{u}_k \longrightarrow 0$ in $\mathcal{S}$, so the continuity of the Fourier transform in $\mathcal{S}$ is proved. Q.E.D.

<u>THEOREM</u> 1.2.3. (Fourier inversion formula). For every u in $\mathcal{S}$,

$$(*) \qquad u(x) = \int e^{2\pi i(x,\xi)} \hat{u}(\xi)\, d\xi.$$

Consequently, the Fourier transform is an isomorphism of $\mathcal{S}$ onto itself.

<u>Proof</u>. We have to calculate the iterated integral

$$\int e^{2\pi i(x,\xi)} \left\{ \int e^{-2\pi i(y,\xi)} u(y)\, dy \right\} d\xi.$$

Unfortunately, we cannot interchange the order of integration because the double integral is not absolutely convergent. To avoid this difficulty, we introduce a "convergence factor" $v(\xi)$, $v \in \mathcal{S}$, to be specified later. This device makes the double integral absolutely convergent, thus interchanging the order of integration we see that

$$\int e^{2\pi i(x,\xi)} v(\xi) \left\{ \int e^{-2\pi i(y,\xi)} u(y)\, dy \right\} d\xi =$$

$$= \int u(y) \left\{ \int e^{-2\pi i(y-x,\xi)} v(\xi)\, d\xi \right\} dy =$$

$$= \int u(y)\, \hat{v}(y - x)\, dy = \int \hat{v}(y)\, u(x + y)\, dy.$$

In brief, we have obtained that

$$(1) \qquad \int e^{2\pi i(x,\xi)} v(\xi)\, \hat{u}(\xi)\, d\xi = \int \hat{v}(y)\, u(x + y)\, dy.$$

It is easy to see that, for any $\varepsilon > 0$, the Fourier transform of $v(\varepsilon\xi)$ is given by $\varepsilon^{-n}\hat{v}(y/\varepsilon)$. So, if in formula (1) we replace $v(\xi)$ by $v(\varepsilon\xi)$, we have

$$(2) \qquad \int e^{2\pi i(x,\xi)} v(\varepsilon\xi)\, \hat{u}(\xi)\, d\xi = \int \hat{v}(y)\, u(x + \varepsilon y)\, dy.$$

Since $\hat{u}$ and $\hat{v}$ are in $\mathcal{S} \subset L^1$ and u and v are bounded and continuous, if we take the limit as $\varepsilon \longrightarrow 0$ in (2) we can pass the limit under the integral sign and obtain

$$(3) \qquad v(0) \int e^{2\pi i(x,\xi)} \hat{u}(\xi)\, d\xi = u(x) \int \hat{v}(y)\, dy.$$

Finally, to obtain (*) from (3) we specify the function v

by taking $v(y) = \exp(-\pi|x|^2)$ and recall (e.g. [6], Chapter II, Example 5) this function (essentially the Weierstrass kernel in E^n) is equal to its own Fourier transform and has integral equal to 1. Q.E.D.

THEOREM 1.2.4. For all u and v in $\mathcal{S}$ we have

(i) $$\int \hat{u}v\,dx = \int u\hat{v}\,dx$$

(ii) $$\int u\bar{v}\,dx = \int \hat{u}\bar{\hat{v}}\,dx \quad \text{(Plancherel formula)}$$

(iii) $$F(u * v)(\xi) = \hat{u}(\xi)\,\hat{v}(\xi)$$

(iv) $$F(uv)(\xi) = (\hat{u} * \hat{v})(\xi).$$

Proof. Letting $x = 0$ in formula (1) of the proof of the preceding theorem we obtain formula (i).

To prove (ii) we set $\bar{\hat{v}} = w$, so that

$$\overline{\hat{w}(\xi)} = \int e^{2\pi i(x,\xi)}\,\hat{v}(x)\,dx$$

and hence $\bar{\hat{w}} = v$ by the Fourier inversion formula. Using formula (i) we see that

$$\int u\bar{v}\,dx = \int u\hat{w}\,dx = \int \hat{u}w\,dx = \int \hat{u}\bar{\hat{v}}\,dx$$

so (ii) is proved.

The steps of the proof of (iii) are as follows:

$$F(u * v)(\xi) = \int e^{-2\pi i(x,\xi)} \left\{\int u(y)\, v(x-y)\, dy\right\} dx =$$

$$= \int e^{-2\pi i(y,\xi)}\, u(y) \left\{\int e^{-2\pi i(x-y,\xi)}\, v(x-y)\, dx\right\} dy =$$

$$= \hat{u}(\xi)\, \hat{v}(\xi).$$

Finally, to prove (iv) it suffices to verify that both sides have the same Fourier transform. Letting $uv = w$ and using the inversion formula we have

$$\hat{\hat{w}}(x) = \int e^{-2\pi i(x,\xi)}\, \hat{w}(\xi)\, d\xi = w(-x) = u(-x)\, v(-x).$$

On the other hand, by (iii) and the inversion formula,

$$F[(\hat{u} * \hat{v})(\xi)]\,(x) = \hat{\hat{u}}(x)\, \hat{\hat{v}}(x) = u(-x)\, v(-x).$$

Q.E.D.

We shall now extend the Fourier transform, by transposition, to the t.v.s. $\mathcal{S}^*$ of temperate distributions.

DEFINITION 1.2.5. The Fourier transform $\hat{f} = Ff$ of a distribution $f \in \mathcal{S}^*$ is defined by

$$\langle \hat{f}, u \rangle = \langle f, \hat{u} \rangle$$

for all $u \in \mathcal{S}$.

From this definition and Theorem 1.2.2 we see that if f is in $\mathcal{S}^*$ so is $\hat{f}$. Moreover formula (i) of Theorem 1.2.4 implies that our new definition agrees with the classical one if f is a function in L^1 or in L^2. In addition, it is easy to verify that $\mathcal{S}^*$ contains all measures $d\mu$ such that for some m

$$\int (1 + |x|)^{-m} |d\mu(x)| < \infty.$$

In particular, for every p, $1 \leq p \leq \infty$, $L^p(E^n) \subset \mathcal{S}^*$. Thus the Fourier transforms of functions in L^p are, in general, temperate distributions.

The Fourier inversion formula can also be written in the form $\hat{u}(x) = u(-x)$ and since $\check{u}(x) = \overline{u(-x)}$ we have the formula $\hat{\hat{u}} = \bar{\check{u}}$. We recall that if g is a distribution so is $\bar{\check{g}}$ and that, by P.1.1.3 and P.1.1.4,

$$\langle \bar{\check{g}}, u \rangle = \overline{\langle \check{g}, \bar{u} \rangle} = \langle g, \check{\bar{u}} \rangle = \langle g, \bar{\check{u}} \rangle.$$

Accordingly, we obtain

<u>THEOREM</u> 1.2.6. The Fourier inversion formula, $\hat{\hat{g}} = \bar{\check{g}}$, is valid for every $g \in \mathcal{S}^*$. Consequently, the Fourier transformation is an isomorphism of $\mathcal{S}^*$ onto itself.

<u>Proof</u>. By 1.2.5 and 1.2.3

$$\langle \hat{\hat{g}}, u \rangle = \langle \hat{g}, \hat{u} \rangle = \langle g, \hat{\hat{u}} \rangle = \langle g, \bar{\check{u}} \rangle = \langle \bar{\check{g}}, u \rangle$$

for all $g \in \mathcal{S}^*$ and $u \in \mathcal{S}$. Thus $\hat{\hat{g}} = \bar{\check{g}}$ in $\mathcal{S}^*$.

Since $u \longrightarrow \hat{u}$ is continuous on $\mathcal{S}$, and $\mathcal{S}^*$ has the weak topology, 1.2.5 clearly implies that the Fourier transformation is also continuous on $\mathcal{S}^*$. Q.E.D.

THEOREM 1.2.7. The Fourier transformation is an isomorphism of L^2 with itself, and an isometry, i.e. we have Parseval formula

$$\int |\hat{g}|^2 \, d\xi = \int |g|^2 \, dx.$$

Proof. For all $g \in L^2$ and $u \in \mathcal{S}$, we have by 1.2.5 and Schwarz's inequality that

$$|\langle \hat{g},u \rangle| = |\langle g,\hat{u} \rangle| = \left| \int g\bar{\hat{u}} \, dx \right| \le \|g\|_2 \, \|\hat{u}\|_2 .$$

Consequently, by Plancherel formula with $u = v$, we obtain

$$|\langle \hat{g},u \rangle| \le \|g\|_2 \, \|u\|_2$$

which shows that $\hat{g}$ defines on L^2 a bounded linear functional, $u \longrightarrow \langle \hat{g},u \rangle$, with norm $\le \|g\|_2$. Hence by Riesz Representation Theorem there is an element of L^2, which we denote again by $\hat{g}$, such that $\|\hat{g}\|_2 \le \|g\|_2$. Applying this last inequality twice we see that

$$\|g\|_2 = \|\hat{\hat{g}}\|_2 \le \|\hat{g}\|_2 \le \|g\|_2$$

so $\|\hat{g}\|_2 = \|g\|_2$ and Parseval formula holds. Q.E.D.

REMARK 1.2.8. Neither our notation nor our terminology about distributions (alias generalized functions) are quite standard. For instance our distributions are conjugate linear functionals, rather than linear functionals as in the more usual terminology. This discrepancy however should cause no trouble and it disappears altogether when the test functions (i.e. elements of $\mathcal{D}$, $\mathcal{S}$, etc.) are real-valued. On the other hand, in Chapter II, we show that a certain linear functional, p.v.k, given by

$$\langle \text{p.v.k}\,, u \rangle = \lim_{\varepsilon \to 0} \int_{|x|>\varepsilon} k(x)\, u(x)\, dx, \qquad u \in \mathcal{D},$$

is a distribution (in the usual sense), meaning of course that this functional is continuous with respect to the topology of $\mathcal{D}$.

EXERCISES

1. Prove that the class $\mathcal{S}$ is closed under convolution.
2. If δ is the Dirac measure, prove that $F(\delta) = \hat{\delta} = 1$. (Hint: Calculate $\langle \hat{\delta}, u \rangle$, $u \in \mathcal{S}$).
3. Prove that if f is a distribution with support at the origin, then $\hat{f}$ is a polynomial. (Hint: Use 1.1.15).
4. Give a different proof of Theorem 1.2.7. (Outline: Plancherel formula in $\mathcal{S}$ implies that Parseval formula holds in $\mathcal{S}$. Thus, if $g_k \in \mathcal{S}$ and $g_k \longrightarrow g$ in L^2 norm, it follows that $\{\hat{g}_k\}$ is a Cauchy sequence in L^2. Clearly $g_k \longrightarrow g$ in $\mathcal{S}^*$, i.e. weakly, hence, by continuity

of the Fourier transformation in $\mathcal{S}^*$, there exists an element of $\mathcal{S}^*$, which we denote by $\hat{g}$, such that $\hat{g}_k \longrightarrow \hat{g}$ in $\mathcal{S}^*$. Show that $\hat{g}$ belongs to L^2 and that $\hat{g}_k \longrightarrow \hat{g}$ in L^2 norm. Then, apply Parseval formula to $\hat{g}_k$ and pass to the limit as $k \longrightarrow \infty$).

CHAPTER II

SINGULAR INTEGRALS AND SOBOLEV SPACES

§1. SINGULAR KERNELS AND THEIR FOURIER TRANSFORMS

We recall that given a function $k(x)$, $x \in E^n$, we defined $k_\varepsilon(x) = \varepsilon^{-n}k(x/\varepsilon)$. If $k(x)$ is locally integrable and $u \in C_o^\infty$ then, for all $\lambda > 0$,

$$\int k(\lambda x)\, \overline{u(x)}\, dx = \lambda^{-n} \int k(x)\, \overline{u(x/\lambda)}\, dx,$$

in other words, $\langle k(\lambda x), u(x)\rangle = \langle k(x), u_\lambda(x)\rangle$. In particular, if $k(x)$ is (positively) homogeneous of degree m, i.e. $k(\lambda x) = \lambda^m k(x)$ for all $\lambda > 0$, we obtain

$$\lambda^m \langle k,u\rangle = \langle k(\lambda x), u(x)\rangle = \langle k(x), u_\lambda(x)\rangle .$$

More generally, for any temperate distribution $f \in \mathcal{S}^*$ we define the operation Π_λ by the formula

(1) $$\langle \Pi_\lambda f, u\rangle = \langle f, u_\lambda\rangle$$

for all $u \in \mathcal{S}$. Clearly then, when f is a locally integrable function $(\Pi_\lambda f)(x) = f(\lambda x)$. It is also clear from (1) that when f belongs to $\mathcal{S}^*$ so does $\Pi_\lambda f$.

Moreover, we have the following extension of the familiar formula for the Fourier transform of $f(\lambda x)$:

$$(2) \qquad F(\Pi_\lambda f) = \lambda^{-n}\Pi_{1/\lambda}F(f).$$

Proof. By 1.2.5 and definition (1) above we see that, for all $u \in \mathcal{S}$,

$$\langle F(\Pi_\lambda f),u\rangle = \langle \Pi_\lambda f,\hat{u}\rangle = \langle f,\lambda^{-n}\Pi_{1/\lambda}\hat{u}\rangle$$

and likewise,

$$\langle \lambda^{-n}\Pi_{1/\lambda}F(f),u\rangle = \langle \hat{f},u(\lambda x)\rangle =$$

$$= \langle f,\lambda^{-n}\Pi_{1/\lambda}\hat{u}\rangle.$$

Q.E.D.

DEFINITION 2.1.0. A distribution f in $\mathcal{S}^*$ is said to be a homogeneous distribution of degree m if $\Pi_\lambda f = \lambda^m f$.

Let us recall that a function f is said to be radial (alias spherically symmetric) if $f(x) = f(|x|)$. Furthermore, we recall (see [6], Chapter II) that the Fourier transform of a radial function is again a radial function.

We begin by considering the locally integrable function

$|x|^{-\alpha}$ where $\frac{n}{2} < \alpha < n$ and $n \geq 2$, and we shall calculate its Fourier transform. Let us split $|x|^{-\alpha}$ into the sum of two functions: $|x|^{-\alpha} = f + g$ where

f = restriction of $|x|^{-\alpha}$ to the set $\{x: |x| \leq 1\}$

$g = |x|^{-\alpha} - f$ = restriction of $|x|^{-\alpha}$ to $\{x: |x| > 1\}$.

Since $\alpha < n$, $f \in L^1$, whereas $\alpha > n/2$ implies that $g \in L^2$. Hence, $F(|x|^{-\alpha}) = \hat{f} + \hat{g}$ is a locally integrable function, being the sum of a bounded continuous function and a function in L^2. Moreover, since $|x|^{-\alpha}$ is locally integrable,

$$\text{(3)} \qquad \Pi_\lambda |x|^{-\alpha} = |\lambda x|^{-\alpha} = \lambda^{-\alpha}|x|^{-\alpha} .$$

Now, formula (2) implies that

$$F(\Pi_\lambda |x|^{-\alpha}) = \lambda^{-n}\Pi_{1/\lambda}F(|x|^{-\alpha})$$

while from (3) it follows that

$$F(\Pi_\lambda |x|^{-\alpha}) = \lambda^{-\alpha}F(|x|^{-\alpha}).$$

Therefore, combining these two identities, we see that

$$F(|x|^{-\alpha}) = \lambda^{\alpha-n}\Pi_{1/\lambda}F(|x|^{-\alpha})$$

or equivalently (replacing $1/\lambda$ by λ).

(4) $$\Pi_\lambda F(|x|^{-\alpha}) = \lambda^{\alpha-n} F(|x|^{-\alpha})$$

so that $F(|x|^{-\alpha})$ is a function homogeneous of degree $\alpha - n$.

In addition $F(|x|^{-\alpha})$ is a radial function because $|x|^{-\alpha}$ is radial. Consequently, being a radial function homogeneous of degree $\alpha - n$, $F(|x|^{-\alpha})$ must be of the form

(5) $$F(|x|^{-\alpha})(\xi) = c_\alpha |\xi|^{\alpha-n}$$

for some complex constant c_α.

The constant c_α may be calculated as follows: recalling that $\exp(-\pi|x|^2)$ belongs to $\mathcal{S}$ and that $F[\exp(-\pi|x|^2)](\xi) = \exp(-\pi|\xi|^2)$, we obtain, using Plancherel formula and (5), that

(6) $$\langle |x|^{-\alpha}, e^{-\pi|x|^2} \rangle = c_\alpha \langle |\xi|^{\alpha-n}, e^{-\pi|\xi|^2} \rangle.$$

Let us introduce the notation $\Sigma = \{x: |x| = 1\}$, W_n = area of Σ, $x' = x/|x|$ = projection of x on Σ, dx' = surface element on Σ. We note also that the change of variable $\pi r^2 = u$, $dr = \frac{1}{2\pi r}\, du = \frac{1}{2}\pi^{-1/2} u^{-1/2} du$, yields,

$$\int_0^\infty e^{-\pi r^2} r^\beta \, dr = \frac{1}{2}\pi^{-(\beta+1)/2} \int_0^\infty e^{-u}\, u^{(\beta-1)/2}\, du.$$

So, in terms of the gamma function $\Gamma(s) = \int_0^\infty e^{-u}\, u^{s-1}\, du$, we obtain the formula

$$(7) \qquad \int_0^\infty e^{-\pi r^2} r^\beta \, dr = \frac{1}{2} \pi^{-(\beta+1)/2} \Gamma\left(\frac{\beta+1}{2}\right)$$

which is valid for all $\beta > -1$.

Now, using polar coordinates $r = |x|$, $dx = r^{n-1} \, dr \, dx'$, we see that

$$\langle |x|^{-\alpha}, e^{-\pi|x|^2} \rangle = \int |x|^{-\alpha} e^{-\pi|x|^2} \, dx =$$

$$= \int_\Sigma dx' \int_0^\infty e^{-\pi r^2} r^{n-1-\alpha} \, dr$$

so, using (7) with $\beta = n - 1 - \alpha$, we obtain

$$(8) \qquad \langle |x|^{-\alpha}, e^{-\pi|x|^2} \rangle = \frac{1}{2} W_n \pi^{-(n-\alpha)/2} \Gamma\left(\frac{n-\alpha}{2}\right) .$$

Similarly, using (7) with $\beta = n - 1 + \alpha - n = \alpha - 1$, we have

$$(8') \qquad c_\alpha \langle |\xi|^{\alpha-n}, e^{-\pi|\xi|^2} \rangle = c_\alpha \frac{1}{2} W_n \pi^{-\alpha/2} \Gamma\left(\frac{\alpha}{2}\right) .$$

Therefore, in view of (6), equating (8) and (8'), we obtain

$$(9) \qquad c_\alpha = \pi^{\alpha-n/2} \Gamma\left(\frac{n-\alpha}{2}\right) \frac{1}{\Gamma\left(\frac{\alpha}{2}\right)} .$$

Summing up, we have proved the following result:

PROPOSITION. 2.1.1. If $n \geq 2$ and $\frac{n}{2} < \alpha < n$, the Fourier transform of the function $|x|^{-\alpha}$ is given by the formula

$$F(|x|^{-\alpha})(\xi) = c_\alpha |\xi|^{\alpha-n}$$

where the constant c_α is given by expression (9) above.

In addition, the argument leading to formula (4) yields the following conclusion:

PROPOSITION. 2.1.2. For any real number α, the Fourier transform of a homogeneous distribution of degree $-\alpha$ is a homogeneous distribution of degree $\alpha - n$.

Let us now pass to the study of certain kernels which are not locally integrable in $E^n, n \geq 1$. According to the notation introduced earlier, we can write a kernel homogeneous of degree $-n$ in the form

$$k(x) = k\left(|x| \frac{x}{|x|}\right) = \Omega(x')|x|^{-n}$$

where $x' = x/|x|$ belongs to the unit sphere Σ and the function Ω, called the characteristic of the kernel k, is the restriction of k to Σ.

Formally, with any such kernel k we associate a distribution p.v.k, called the principal value of k, given by

$$\langle p.v.k, f\rangle = \lim_{\varepsilon\to o} \int_{|x|>\varepsilon} k(x)\, f(x)\, dx$$

and an operator K, of convolution type, given by

$$(*)\quad \begin{cases} (Kf)(x) = p.v. \int k(x-y)\, f(y)\, dy = \\ \quad = \lim_{\varepsilon\to o} \int_{|x-y|>\varepsilon} k(x-y)\, f(y)\, dy = \\ \quad = \lim_{\varepsilon\to o} (k_\varepsilon * f)(x),\ o < \varepsilon < 1, \end{cases}$$

where we have set $k_\varepsilon(x) = \begin{cases} k(x) & \text{if } |x| > \varepsilon \\ o & \text{otherwise.} \end{cases}$

We note in passing that, in view of the truncations, the kernels k_ε are locally integrable if the characteristic Ω is integrable on Σ.

In order for these operators to exist for a sufficiently wide class of functions f and to be bounded operators on such classes, we shall impose the following conditions on the singular kernels:

DEFINITION. 2.1.3. A function $k(x) = \Omega(x')|x|^{-n}$ is said to be a singular convolution-kernel if

(a) k has mean value zero on Σ, i.e.

$$\int_\Sigma k = \int_\Sigma \Omega(x')\, dx' = 0$$

(b) $\Omega \in L^q(\Sigma)$ for some $q > 1$.

The operator K associated by (*) to a singular convolution-kernel is said tc be a <u>singular</u> <u>convolution-operator</u>.

Let us write $\|\Omega\|_q = \left\{\int_\Sigma |\Omega(x')|^q \, dx'\right\}^{1/q}$.

The following result about singular convolution-operators constitutes the starting point for the theory of singular integral operators of Calderón and Zygmund which will be presented in the rest of these Notes.

<u>THEOREM</u>. 2.1.4. Let $f \in L^p(E^n)$, $1 < p < \infty$, and set $\tilde{f}_\varepsilon = k_\varepsilon * f$ where k_ε is the truncation of a singular convolution-kernel k. Then, for some constant $A > 0$ independent of f and ε,

(i) $\|\tilde{f}_\varepsilon\|_p \leq A\|f\|_p$, where $A = A_{p,q}\|\Omega\|_q$ and $A_{p,q}$ depends only on p, q and the dimension n.

(ii) As $\varepsilon \to 0$ sequentially, $\{\tilde{f}_\varepsilon\}$ is a Cauchy sequence in L^p.

(iii) If $\tilde{f} = Kf$ denotes the limit in L^p norm of $\{\tilde{f}_\varepsilon\}$,

$$\|\tilde{f}\|_p \leq A\|f\|_p \ .$$

The proof of this theorem (see [6], Chapter IV, Theorem 7) is rather deep and lengthy and will not be repeated here.

It consists of two main steps: the reduction of the general case to an odd kernel $k(-x) = -k(x)$, and the reduction of this latter case to the one-dimensional singular kernel $k(x) = \frac{1}{x}$ of the Hilbert transform ([6], Chapter III). In essence, Theorem 2.1.4 establishes the existence of singular convolution operators as limits in L^p norm, $1 < p < \infty$, of the truncated convolutions $\tilde{f}_\varepsilon = k_\varepsilon * f$. Furthermore, it shows that these operators $f \longrightarrow Kf$ define bounded linear transformations in $L^p(E^n)$.

<u>LEMMA</u>. 2.1.5. Let $k(x) = \Omega(x')|x|^{-n}$ be a singular convolution-kernel. Then the linear functional p.v.k given by

$$\langle \text{p.v.}k,u \rangle = \lim_{\varepsilon \to 0} \int_{|x|>\varepsilon} k(x)\, u(x)\, dx, \quad u \in \mathcal{D},$$

exists and is a distribution.

<u>Proof</u>. With $u \in \mathcal{D}$ let us write

$$\int_{|x|>\varepsilon} k(x)\, u(x)\, dx = \int_{1>|x|>\varepsilon} + \int_{|x|\geq 1} = I + II .$$

By Hölder's inequality with $\frac{1}{q} + \frac{1}{q'} = 1$, we see that

$$|II| = \left| \int_{|x|\geq 1} k(x)\, u(x)\, dx \right| \leq \left(\int_{|x|\geq 1} |k(x)|^q\, dx \right)^{1/q} \|u\|_{q'}$$

and with $\alpha = nq - n > 0$ since $q > 1$ it follows that

$$\int_{|x|\geq 1} |k(x)|^q \, dx = \int_{\Sigma} |\Omega(x')|^q \, dx' \int_1^{\infty} \frac{dr}{r^{1+\alpha}} < \infty$$

since $\Omega \in L^q(\Sigma)$. Consequently, II is finite.

By the mean value zero condition of Ω,

$$\int_{1>|x|>\varepsilon} k(x) \, dx = \int_{\Sigma} \Omega(x') \, dx' \int_{\varepsilon}^{1} \frac{dr}{r} = o$$

so we can write

$$I = \int_{1>|x|>\varepsilon} k(x)[u(x) - u(o)] \, dx = \int_{1>|x|>\varepsilon} g(x) \, dx$$

and we can let $\varepsilon \to o$ since the integral is convergent because $g(x)$ is absolutely integrable on the unit ball. In fact, if C is a bound for the first order derivatives of u, $|g(x)| \leq$ $\leq C|\Omega(x')| \; |x|^{1-n}$. Therefore,

$$\int_{|x|\leq 1} |g(x)| \, dx \leq C \int_{\Sigma} |\Omega(x')| \, dx' \int_{o}^{1} dr < \infty$$

since $\Omega \in L^q(\Sigma) \subset L^1(\Sigma)$.

Hence, letting $k_{\varepsilon}(x) = k(x)$ if $|x| > \varepsilon$ and $k_{\varepsilon}(x) = 0$ otherwise, we have proved that, as $\varepsilon \to 0$, $\langle k_{\varepsilon}, u\rangle \to \langle p.v.k, u\rangle$ for every $u \in \mathcal{D}$.

Let us write $\|\Omega\|_1 = \int_{\Sigma} |\Omega(x')| \, dx'$ and, for some $0 < \theta < 1$,

$$u(x) - u(o) = \sum_{j=1}^{n} x_j \frac{\partial u}{\partial x_j}(\theta x) .$$

If (supp u) $\subset$ K, K compact, we have

$$|\langle k_\varepsilon, u\rangle| = \left| \int_\varepsilon^\infty \int_\Sigma \Omega(x')[u(x) - u(o)]\, r^{-1}\, dr\, dx' \right|$$

and since $\left|\frac{x_j}{|x|}\right| \leq 1$ we obtain, for some constant C_K depending on K,

$$|\langle k_\varepsilon, u\rangle| \leq C_K \|\Omega\|_1 \sum_{j=1}^{n} \sup_x \left|\frac{\partial u}{\partial x_j}\right| .$$

Therefore, letting $\varepsilon \to 0$, we conclude that

$$|\langle p.v.k, u\rangle| \leq C_K \|\Omega\|_1 \sum_{j=1}^{n} \sup_x \left|\frac{\partial u}{\partial x_j}\right| .$$

From this estimate it is clear that if $\{u_m\}$ converges to 0 in $\mathcal{D}$, as $m \to \infty$, then $\langle p.v.k, u_m\rangle \to 0$. Consequently, p.v.k is a distribution.

Q.E.D.

The next theorem yields some important additional information on p.v.k, and on its Fourier transform.

THEOREM. 2.1.6. If $k(x) = \Omega(x')|x|^{-n}$ is a singular convolution-kernel then p.v.k(x) is a temperate distribution and F(p.v.k) is a bounded function homogeneous of degree zero with mean value zero on Σ. In addition, for all $f \in \mathcal{D}$,

(i) $Kf = (p.v.k) * f$

(ii) $F(Kf) = F(p.v.k)\, F(f)$.

<u>Proof</u>. It suffices to show that $\ell = (\text{p.v.}k)(1+|x|^2)^{-1}$ is a temperate distribution. In fact, if this is so, then clearly $(\text{p.v.}k) = \ell(1+|x|^2)$ is again a temperate distribution. Writing ℓ in the form $\ell = \ell_1 + \ell_2$, where ℓ_1 = restriction of ℓ to $\{x:|x| < 1\}$ and $\ell_2 = \ell - \ell_1$ = restriction of ℓ to $\{x:|x| \geq 1\}$, we see that ℓ_1 is a distribution with compact support, hence temperate since $\mathcal{D}_o^* \subset \mathcal{S}^*$. On the other hand, for $|x| \geq 1$, $\ell_2 = k(x)(1 + |x|^2)^{-1}$, and using polar coordinates

$$\int_{|x|\geq 1} |k(x)|(1 + |x|^2)^{-1}\, dx = \int_\Sigma |\Omega(x')|\, dx' \int_1^\infty \frac{dr}{r+r^3} < \infty$$

since $\Omega \in L^q(\Sigma) \subset L^1(\Sigma)$ for $q > 1$. In other words, ℓ_2 coincides with an integrable function and hence is also a temperate distribution. Consequently, (p.v.k) is a temperate distribution.

Formula (i) is a direct consequence of the definition (*) of Kf and of the definition of convolution of elements of $\mathcal{S}^* \subset \mathcal{D}^*$ with functions in $\mathcal{D}$.

The implication (i) $\Rightarrow$ (ii) is not difficult if we take into account the results of Chapter I and formula (iii) of Theorem 1.2.4. At any rate, it constitutes a good exercise for the reader.

Now it remains to verify that $F(p.v.k)$ has the stated properties. If $f \in \mathcal{D} \subset L^2$ then, using Theorem 2.1.4 with $p = 2$, we have that $Kf \in L^2$ and hence by (ii) $F(Kf) = F(p.v.k)\, F(f)$ is also in L^2.

Given any compact set $C \subset E^n$ let us choose an $f \in \mathcal{D}$ such that $F(f)$ never vanishes and $F(f)\, F(p.v.k) = F(Kf)$ on C. Then, since $F(f)\, F(p.v.k) \in L^2$ and $1/F(f)$ is bounded on C, it follows that $F(p.v.k) \in L^2(C)$. Hence $F(p.v.k)$ coincides with a function which is locally in L^2.

From (ii), Parseval formula in L^2 and Theorem 2.1.4 we deduce that, for all $f \in \mathcal{D}$,

$$\|F(p.v.k)\, F(f)\|_2 = \|F(Kf)\|_2 = \|Kf\|_2 \leq A\|f\|_2$$

therefore, for all $f \in \mathcal{D}$,

$$(1) \qquad \|F(p.v.k)\, F(f)\|_2 \leq A\|F(f)\|_2 .$$

But, since $F(\mathcal{D})$ is dense in L^2, formula (1) implies that $F(p.v.k)$ must belong to L^∞.

Since $k(x)$ is homogeneous of degree $-n$, it is easy to check that $p.v.k$ is a homogeneous distribution of degree $-n$.

Therefore, by Proposition 2.1.2, we deduce that F(p.v.k) is a homogeneous distribution of degree zero, i.e. for all $\lambda > 0$

$$\Pi_\lambda F(p.v.k) = F(p.v.k) .$$

But F(p.v.k) is a function in L^∞, hence locally integrable, so it follows that for all $u \in \mathcal{D}$

$$\langle F(p.v.k), u \rangle = \langle \Pi_\lambda F(p.v.k), u \rangle =$$

$$= \langle F(p.v.k), u_\lambda \rangle =$$

$$= \langle F(p.v.k)(\lambda x), u \rangle .$$

Therefore, for each $\lambda > 0$, we conclude that for almost every x

$$(2) \qquad F(p.v.k)(\lambda x) = F(p.v.k)(x) .$$

Let us note here that the set of points x for which (2) does not hold depends on λ.

<u>CLAIM</u>: We can modify the function F(p.v.k) on a set of measure zero in such a way that (2) will hold for all $x \neq 0$ and all $\lambda > 0$.

Proof of the Claim. Letting $h = F(p.v.k)$ and using the fact that $h(x)$ is measurable, it follows from (2) that the equality $h(\lambda x) = h(x)$ is valid for almost every point (x,λ) in $E^n X(0,\infty)$.

Denote by Z the set of points x for which $h(\lambda x) \neq h(x)$ on a set of values of λ having positive linear measure. Since $h(\lambda x) = h(x)$ almost everywhere on $E^n X(0,\infty)$, Z must have measure zero. But then it follows from Fubini's theorem that, for some $\rho > 0$, the intersection of the sphere $\Sigma_\rho = \{x: |x| = \rho\}$ with Z has zero surface measure. With this fixed ρ we define

$$h^*(x) = \begin{cases} h\left(\rho \frac{x}{|x|}\right) & \text{if } x \neq 0 \text{ and } \rho\frac{x}{|x|} \notin Z \\ 0 & \text{otherwise.} \end{cases}$$

Clearly $h^*(x)$ is measurable and homogeneous of degree zero. So, it remains to verify that $h^* = h$ almost everywhere.

Let $x \neq 0$ so $x_\rho = \rho\frac{x}{|x|} \in \Sigma_\rho$. If $x_\rho \notin Z$ then $h^*(x_\rho) = h(x_\rho)$ therefore, since $\Sigma_\rho \cap Z$ has zero surface measure, $h^* = h$ almost everywhere on Σ. Moreover, for all $\lambda > 0$, $h^*(\lambda x_\rho) = h^*(x_\rho) = h(x_\rho)$, and, since $h(x_\rho) = h(\lambda x_\rho)$ for almost every $\lambda > 0$, we conclude that $h^*(x) = h(x)$ almost everywhere. Hence the Claim is proved.

Thus, after the modification carried out in the preceding Claim, we have established that $F(p.v.k)$ is a bounded function, homogeneous of degree zero. It remains to show that it has mean value zero on the unit sphere Σ. Now, by definition of F on $\mathcal{S}^*$,

$$\langle F(p.v.k), e^{-\pi|\xi|^2} \rangle = \langle p.v.k, e^{-\pi|x|^2} \rangle$$

so,

$$\int F(\text{p.v.}k)(\xi)\ e^{\pi|\xi|^2}\ d\xi = \lim_{\varepsilon\to 0} \int_{|x|>\varepsilon} k(x)\ e^{-\pi|x|^2}\ dx\ .$$

But

$$\int_{|x|>\varepsilon} k(x)\ e^{-\pi|x|^2}\ dx = \int_\Sigma \Omega(x')\ dx' \int_\varepsilon^\infty r^{-1}e^{-\pi r^2}\ dr = 0$$

because Ω has mean value zero on Σ. Consequently,

$$(3) \qquad \int F(\text{p.v.}k)(\xi)\ e^{-\pi|\xi|^2} d\xi = 0.$$

On the other hand, with $\rho = |\xi|$ and $\xi' = \xi/|\xi|$,

$$\int F(\text{p.v.}k)(\xi)\ e^{-\pi|\xi|^2}\ d\xi = \int_\Sigma F(\text{p.v.}k)(\xi')\ d\xi' \int_0^\infty e^{-\pi\rho^2}\ \rho^{n-1}\ d\rho$$

where the last integral is positive since the integrand is a positive integrable function. Therefore, in view of (3), we must have that

$$\int_\Sigma F(\text{p.v.}k) = 0\ .$$

This completes the proof of the theorem.

Q.E.D.

We denote by $E_o^n = E^n - \{0\}$ the real Euclidean n-space with the origin removed.

<u>THEOREM</u>. 2.1.7. Let $k \in C^{\infty}(E_o^n)$ be homogeneous of degree $-n$ and with mean value zero on Σ. Then, $F(p.v.k) = h$ belongs to $C^{\infty}(E_o^n)$, is homogeneous of degree zero and has mean value zero on Σ.

Conversely, if $h \in C^{\infty}(E_o^n)$ is homogeneous of degree zero and has mean value zero on Σ, then $h = F(p.v.k)$ for some k as above.

<u>Proof</u>. By Theorem 2.1.6, h is a bounded function, homogeneous of degree zero and with mean value zero on Σ. So, to obtain the first conclusion, it remains to prove that $h \in C^{\infty}(E_o^n)$.

It is easy to see that, for any temperate distribution T, $F\left(\frac{\partial}{\partial x_j}[x_i T]\right) = -x_j \frac{\partial}{\partial x_i} F(T)$. Consequently,

$$(1) \qquad -x_j \frac{\partial}{\partial x_i} F(p.v.k) = F\left(\frac{\partial}{\partial x_j}[x_i (p.v.k)]\right) .$$

Now, for all $u \in \mathcal{D}$, assuming for simplicity that $u = \bar{u}$ is real,

$$\left\langle \frac{\partial}{\partial x_j}[x_i (p.v.k)] , u \right\rangle = -\left\langle p.v.k, x_i \frac{\partial u}{\partial x_j} \right\rangle =$$

$$= - \lim_{\varepsilon \to 0} \int_{|x|>\varepsilon} k(x)\, x_i \frac{\partial u}{\partial x_j}\, dx .$$

Since $x_i k \frac{\partial u}{\partial x_j} + \left(\frac{\partial}{\partial x_j}[x_i k]\right) u = \frac{\partial}{\partial x_j}[(x_i k)\, u]$, if $(\operatorname{supp} u) \subset \{x : |x| < R\}$ then, applying Gauss formula on the domain $\varepsilon < |x| < R$, we see that

$$\int_{|x|>\varepsilon} \frac{\partial}{\partial x_j} [(x_i k)\ u]\ dx = - \int_{|x|=\varepsilon} x_i k(x)\ u(x) \frac{x_j}{|x|}\ d\sigma_\varepsilon$$

where $d\sigma_\varepsilon$ is the surface element on the sphere $|x| = \varepsilon$, and therefore

$$- \int_{|x|>\varepsilon} k(x)\ x_i \frac{\partial u}{\partial x_j}\ dx = \int_{|x|>\varepsilon} \left(\frac{\partial}{\partial x_j} [x_i k(x)]\right) u(x)\ dx +$$

$$+ \int_{|x|=\varepsilon} x_i k(x)\ u(x) \frac{x_j}{|x|}\ d\sigma_\varepsilon \quad .$$

Letting $\varepsilon \to 0$, the last integral tends to $Cu(0)$, for some constant C. Hence we have obtained the formula

$$(2) \qquad \frac{\partial}{\partial x_j} [x_i (\text{p.v.}k)] = \text{p.v.} \frac{\partial}{\partial x_j} [x_i k] + C\delta$$

where δ is the Dirac measure. Since $F(\delta) = 1$ (cf. Exercise 2 at the end of Chapter I), taking the Fourier transform in (2) and using (1) we conclude that

$$(3) \qquad -x_j \frac{\partial}{\partial x_i} F(\text{p.v.}k) = F(\text{p.v.} \frac{\partial}{\partial x_j} [x_i k]) + C \ .$$

Since $k \in C^\infty(E_o^n)$ and is homogeneous of degree $-n$, clearly the function $\frac{\partial}{\partial x_j} [x_i k]$ also has these properties. Moreover, we shall see next that this function, like k itself, has mean value zero on the unit sphere. In fact, using polar coordinates we see that

$$\int_{1\le|x|\le 2}\left(\frac{\partial}{\partial x_j}[x_i k]\right)(x)\,dx = (\log 2)\int_\Sigma \frac{\partial}{\partial x_j}[x_i k]\,dx'$$

whereas, by Gauss formula, the first integral equals

$$\int_{|x|=2} x_i k(x)\frac{x_j}{|x|}\,d\sigma_2 - \int_{|x|=1} x_i k(x)\frac{x_j}{|x|}\,d\sigma_1 = 0$$

because both integrals here are equal since the integrand is homogeneous of degree $1-n$. Therefore,

$$\int_\Sigma \frac{\partial}{\partial x_j}[x_i k]\,dx' = 0 .$$

Now, by Theorem 2.1.6, we can conclude that $F\left(\text{p.v.}\frac{\partial}{\partial x_j}[x_i k]\right)$ is a bounded function. But then, by (3), it follows that $\frac{\partial}{\partial x_i}F(\text{p.v.}k)$ coincides with a bounded function over all compact sets for which $x_j \neq 0$. However, since this must hold for each $j = 1,\ldots,n$, we must have that $\frac{\partial}{\partial x_i}F(\text{p.v.}k)$ coincides with a locally bounded function on

$$E_o^n = E^n - \{0\} .$$

Repeating this argument we see that for each multi-index α, $|\alpha| > 0$, there is a constant C_α such that

$$\frac{\partial^{|\alpha|}}{\partial x_j^{|\alpha|}}[x^\alpha(\text{p.v.}k)] = \text{p.v.}\frac{\partial^{|\alpha|}}{\partial x_j^{|\alpha|}}[x^\alpha k] + C_\alpha\delta$$

and

$$(-1)^{|\alpha|} x_j^{|\alpha|} \left(\frac{\partial}{\partial x}\right)^{\alpha} F(p.v.k) = F\left(\frac{\partial^{|\alpha|}}{\partial x_j^{|\alpha|}} [x^{\alpha}(p.v.k)]\right) =$$

$$= F\left(p.v. \frac{\partial^{|\alpha|}}{\partial x_j^{|\alpha|}} [x^{\alpha}k]\right) + C_{\alpha} .$$

Then, arguing as before, we note that $\frac{\partial^{|\alpha|}}{\partial x_j^{|\alpha|}} [x^{\alpha}k]$ is again a kernel satisfying the assumptions of Theorem 2.1.6 and hence $\left(\frac{\partial}{\partial x}\right)^{\alpha} F(p.v.k)$ coincides with a function which is bounded on every compact subset of E_o^n. Therefore we can conclude that $F(p.v.k) \in C^{\infty}(E_o^n)$ and since its derivatives have negative degrees of homogeneity they must be bounded on the complement of any open neighborhood of the origin.

Conversely, suppose that $h \in C^{\infty}(E_o^n)$ is homogeneous of degree zero and has mean value zero on Σ. We must find a function G, having the same properties as k above, such that $F(p.v.G) = h$.

Since h is also a temperate distribution, homogeneous of degree zero, there exists a temperate distribution H, homogeneous of degree $-n$, such that $F(H) = h$. Then, for each $j = 1,\ldots n$, it is easy to see that

$$\frac{\partial^n}{\partial x_j^{\,n}} h = (2\pi i)^n F(\xi_j^n H) . \tag{4}$$

On the other hand, $\frac{\partial^n}{\partial x_j^{\,n}} h \in C^{\infty}(E_o^n)$, is homogeneous of degree -n and, writing it in the form

$$\frac{\partial}{\partial x_j}\left[\frac{\partial^{n-1}}{\partial x_j^{\,n-1}} h\right]$$

and integrating over $1 \leq |x| \leq 2$, we see by an argument used earlier in this proof that it has mean value zero on Σ. Thus, by 2.1.5 and 2.1.6,

$$\text{p.v.} \frac{\partial^n}{\partial x_j^{\,n}} h$$

is a temperate distribution. Furthermore, we have that

$$\frac{\partial^n}{\partial x_j^{\,n}} h \quad \text{and} \quad \text{p.v.} \frac{\partial^n}{\partial x_j^{\,n}} h$$

differ by a distribution with support at the origin. Consequently, by 1.1.15 we have that, for some $m \geq 0$,

$$(4') \qquad \frac{\partial^n}{\partial x_j^{\,n}} h = \text{p.v.} \frac{\partial^n}{\partial x_j^{\,n}} h + \sum_{|\alpha| \leq m} c_\alpha D^\alpha \delta .$$

Taking inverse Fourier transforms in (4) and (4') and comparing the results, we obtain

$$(5) \qquad (2\pi i)^n \xi_j^n H = F^{-1}\left(\text{p.v.} \frac{\partial^n}{\partial x_j^{\,n}} h\right) + \sum_{|\alpha| \leq m} c_\alpha \xi^\alpha .$$

But the first term of the right side of (5), as the inverse Fourier transform of a temperate distribution homogeneous of degree -n, is homogeneous of degree zero.

Since $\xi_j^{\,n} H$ is also homogeneous of degree zero, the polynomial in (5) must be a constant, so that

$$(5') \qquad (2\pi i)^n \xi_j^{\,n} H = F^{-1}\left(\text{p.v.}\ \frac{\partial^n}{\partial x_j^{\,n}} h\right) + c_o .$$

Moreover, since

$$\frac{\partial^n}{\partial x_j^{\,n}} h \in C^\infty (E_o^{\,n})$$

and is homogeneous of degree $-n$ with mean value zero on Σ and since F and F^{-1} have identical properties, it follows from (5') and the part of this theorem which has already been proved that $\xi_j^{\,n}H$ belongs to $C^\infty(E_o^{\,n})$ and is a bounded function homogeneous of degree zero. Hence, dividing by $\xi_j^{\,n}$ and recalling that $j = 1,\dots,n$ is arbitrary, we conclude that H coincides on $E_o^{\,n}$ with a function $G \in C^\infty(E_o^{\,n})$, which is homogeneous of degree $-n$. We verify next that G has mean value zero on Σ.

Let $u \in \mathcal{D}$ be a radial function which is positive on $1 < |x| < 2$, say, and vanishes elsewhere. Then, by definition of G, we see that

$$\langle H, u \rangle = \int G(x)\, u(x)\, dx = C \int_\Sigma G(x')\, dx'$$

where $C = \int_1^2 u(r)\, r^{-1}\, dr > 0$. On the other hand, since

$H = F^{-1}(h)$ and h is homogeneous of degree zero,

$$\langle H, u \rangle = \langle F^{-1}(h), u \rangle = \langle h, F^{-1}(u) \rangle =$$

$$= \int_0^\infty [F^{-1}u](r)\, r^{n-1}\, dr \int_\Sigma h(x')\, dx' = 0$$

because $F^{-1}(u)$ is again a radial function and h has mean value zero on Σ. Therefore, G has mean value zero on Σ.

Finally, since the distribution H - p.v.G has support at the origin, its Fourier transform h - F(p.v.G) is a polynomial which must be a constant because both h and F(p.v.G) are bounded functions. Moreover, this constant must be zero since both h and F(p.v.G) have mean value zero on Σ. Consequently, h = F(p.v.G) where G has all the required properties.

Q.E.D.

§2. THE SOBOLEV SPACES L_k^p AND GENERALIZATIONS

Let us recall the definition of the function spaces $L^p = L^p(E^n)$. We consider the set of (Lebesque) measurable functions f such that, for $0 < p \leq \infty$, the quantity

$$\|f\|_p = \begin{cases} \left(\int |f|^p \, dx\right)^{1/p}, & \text{if } 0 < p < \infty \\ \text{essential supremum of } |f|, & \text{if } p = \infty \end{cases}$$

is finite. Identifying those functions which coincide almost everywhere, we have that if $1 \leq p \leq \infty$ then the linear spaces $L^p = \{f : \|f\|_p < \infty\}$ are Banach spaces (i.e. complete normed linear spaces) with respect to the norm $\|f\|_p$. Moreover, by Hölder's inequality, it follows that, for $1 \leq p \leq \infty$, $L^p \subset L^1_{loc}$, that is, functions in L^p are locally integrable in E^n.

Given a function $K(x) \in L^1$ such that $\int K(x)\,dx = 1$ and setting $K_\varepsilon(x) = \varepsilon^{-n} K(x/\varepsilon)$, $\varepsilon > 0$, it is easy to see that $K_\varepsilon(x)$ satisfies the following properties:

a) for all ε, $\|K_\varepsilon\|_1 = \|K\|_1$

b) $\int K_\varepsilon(x)\,dx = 1$

c) for any $a > 0$, $\int_{|x|>a} |K_\varepsilon(x)|\,dx \to 0$, as $\varepsilon \to 0$.

Accordingly, if $f \in L^p$, where $1 \leq p < \infty$, the convolutions

$$(f * K_\varepsilon)(x) = \int f(y)\, K_\varepsilon(x - y)\, dy$$

converge to f in L^p norm, as $\varepsilon \to 0$. (See [6], Chapter I, Theorem 5).

Furthermore, the functions in L^p with compact support are dense in L^p if $1 \leq p < \infty$, and one can verify without difficulty that, for $f \in L^p$ with compact support and $g \in \mathcal{D}$, the convolution $f * g$ also belongs to $\mathcal{D}$. Consequently, from the preceding facts, one readily obtains the following result.

<u>PROPOSITION</u>. 2.2.0. Let $f \in L^p$, $1 \leq p < \infty$, and let $K(x) \in \mathcal{D}$ satisfy $\int K(x)\, dx = 1$. If $K_\varepsilon(x) = \varepsilon^{-n}K(x/\varepsilon)$ then the convolutions $f * K_\varepsilon$ converge to f in L^p norm as $\varepsilon \to 0$. Furthermore, $\mathcal{D}$ is dense in L^p for every finite $p \geq 1$.

In Chapter I we saw that, for all $p \geq 1$, L^p can also be viewed as a subset of $\mathcal{S}^*$, the space of all temperate distributions; hence functions in L^p are differentiable in the sense of distributions. So we are led to consider those functions $f \in L^p$ whose distribution derivatives $D^\alpha f$ also belong to L^p in the following sense: there exist functions $g_\alpha \in L^p$ such that

$$\langle D^\alpha f, u\rangle = \langle f, D^\alpha u\rangle = \langle g_\alpha, u\rangle$$

for all $u \in \mathcal{D}$, and thus, as distributions, $D^\alpha f = g_\alpha \in L^p$.

<u>DEFINITION</u>. 2.2.1. Let $1 \leq p \leq \infty$ and let k be a non-negative integer. Then $L^p_k = L^p_k(E^n)$ is defined to be the class of all functions $f \in L^p$ whose distribution derivatives $D^\alpha f$ also belong to L^p for every multi-index α such that $|\alpha| \leq k$.

The class L_k^p will be equipped with the norm

$$\|f\|_{p,k} = \left\{ \sum_{|\alpha| \leq k} \|D^\alpha f\|_p^2 \right\}^{1/2} .$$

REMARK. Clearly, L_k^p is a linear submanifold of L^p and $L_o^p = L^p$. Also it is easy to see that linear functional $\|f\|_{p,k}$ defined above is indeed a norm on the linear space L_k^p. The functionals

$$\sum_{|\alpha| \leq k} \|D^\alpha f\|_p \quad \text{and} \quad \sup_{|\alpha| \leq k} \|D^\alpha f\|_p$$

yield equivalent norms on the class L_k^p.

Let us note that, for every $0 \leq j \leq k$, $L_k^p \subset L_j^p \subset L^p$ where the inclusions are continuous since, for all $f \in L_k^p$, $\|f\|_p \leq \|f\|_{p,j} \leq \|f\|_{p,k}$.

THEOREM. 2.2.2. For all $p \geq 1$ and positive integers k, L_k^p is a Banach space. Moreover, if p is finite, $\mathcal{D}$ is dense in L_k^p.

Proof. To establish the first conclusion we have to show that L_k^p is complete. Let $\{f_n\}$ be a Cauchy sequence in L_k^p. Then, by definition of the norm, we see that for each α, with $0 \leq |\alpha| \leq k$, $\{D^\alpha f_n\}$ is a Cauchy sequence in L^p. In particular, $\{f_n\}$ is a Cauchy sequence in L^p, so, for some $g_o \in L^p$, $f_n \rightarrow g_o$ in L^p norm.

Similarly, for each α with $0 < |\alpha| \leq k$, there exists a function $g_\alpha \in L^p$ such that $D^\alpha f_n \to g_\alpha$ in L^p.

Let us recall that, for all $p \geq 1$, if $h_n \to h$ in L^p then, a fortiori $\langle h_n, u\rangle \to \langle h, u\rangle$ for all $u \in \mathcal{D}$; in fact, by Hölder's inequality, $|\langle h_n - h, u\rangle| \leq \|h_n - h\|_p \|u\|_q \to 0$, where q is the conjugate exponent of p. Hence, from the preceding paragraph, it follows that for all $u \in \mathcal{D}$

$$\langle D^\alpha f_n, u\rangle \to \langle g_\alpha, u\rangle \quad \text{and}$$

$$\langle D^\alpha f_n, u\rangle = \langle f_n, D^\alpha u\rangle \to \langle g_o, D^\alpha u\rangle = \langle D^\alpha g_o, u\rangle .$$

Consequently, $D^\alpha g_o = g_\alpha \in L^p$ for all $|\alpha| \leq k$, so $g_o \in L^p_k$ and $f_n \to g_o$ in L^p_k norm.

Now, with $1 \leq p < \infty$, we shall prove that $\mathcal{D}$ is dense in L^p_k. Given $f \in L^p_k$, let us consider the convolutions

$$h_\varepsilon(x) = (f * K_\varepsilon)(x) = \int f(y)\, K_\varepsilon(x - y)\, dy$$

where, as usual, $K_\varepsilon(x) = \varepsilon^{-n} K(x/\varepsilon)$, $K(x) \in \mathcal{D}$ and $\int K(x)\, dx = 1$. By Proposition 2.2.0, we know that $h_\varepsilon \to f$ in L^p norm, as $\varepsilon \to 0$. Since, for every multi-index α, $D^\alpha h_\varepsilon = D^\alpha(f * K_\varepsilon) = f * D^\alpha K_\varepsilon$, it follows easily that $h_\varepsilon(x) \in C^\infty$.

Moreover,

$$\left(\frac{\partial}{\partial x}\right)^{\alpha} h_{\varepsilon}(x) = \int f(y) \left(\frac{\partial}{\partial x}\right)^{\alpha} K_{\varepsilon}(x-y)\, dy =$$

$$= (-1)^{|\alpha|} \int f(y)\left(\frac{\partial}{\partial y}\right)^{\alpha} K_{\varepsilon}(x-y)\, dy$$

so, if $|\alpha| \leq k$, integrating by parts we obtain

$$(*) \quad D^{\alpha}h_{\varepsilon}(x) = \int [D^{\alpha}f](y)\, K_{\varepsilon}(x-y)\, dy = [(D^{\alpha}f) * K_{\varepsilon}](x)$$

where $D^{\alpha}f \in L^{p}$. Hence, by Young's Theorem,

$$\|D^{\alpha}h_{\varepsilon}\|_{p} \leq \|D^{\alpha}f\|_{p}\ \|K_{\varepsilon}\|_{1} = A\|D^{\alpha}f\|_{p}$$

since $\|K_{\varepsilon}\|_{1} = \|K\|_{1} = A$. Consequently, for all $\varepsilon > 0$, $h_{\varepsilon} \in L^{p}_{k}$ and, applying 2.2.0 to formula (*), we also see that for all $|\alpha| \leq k$, $D^{\alpha}h_{\varepsilon} \longrightarrow D^{\alpha}f$ in L^{p} norm, as $\varepsilon \longrightarrow 0$. Therefore, $h_{\varepsilon}(x) \in C^{\infty}$ and $h_{\varepsilon} \longrightarrow f$ in L^{p}_{k} norm, as $\varepsilon \longrightarrow 0$.

Finally, choosing $u(x) \geq 0$ in $\mathcal{D}$, such that $u(x) = 1$ if $|x| \leq 1$, we consider the functions

$$g_{\varepsilon}(x) = u(\varepsilon x)\, h_{\varepsilon}(x)\ .$$

Then,

$$D^{\alpha}[g_{\varepsilon}(x) - h_{\varepsilon}(x)] = D^{\alpha}(h_{\varepsilon}(x)[u(\varepsilon x) -1]) =$$

$$= \sum_{\substack{|\gamma|+|\beta|=|\alpha| \\ |\gamma|\geq 1}} C_{\beta\gamma} D^{\beta}h_{\varepsilon}(x)\, D^{\gamma}u(\varepsilon x)\, \varepsilon^{|\gamma|} +$$

$$+ [u(\varepsilon x) -1]\, D^{\alpha}h_{\varepsilon}(x)$$

and, since for all $0 \leq |\alpha| \leq k$ and all $\varepsilon > 0$, $\|D^{\alpha}h_{\varepsilon}\|_p \leq A\|D^{\alpha}f\|_p$ with A independent of ε, it follows that

$$\|D^{\alpha}g_{\varepsilon} - D^{\alpha}h_{\varepsilon}\|_p \longrightarrow 0, \quad \text{as } \varepsilon \longrightarrow 0.$$

Consequently, $(g_{\varepsilon}-h_{\varepsilon}) \longrightarrow 0$ in L^p_k norm. Hence $g_{\varepsilon} \in \mathcal{D}$ and $g_{\varepsilon} \longrightarrow f$ in L^p_k norm, as $\varepsilon \longrightarrow 0$.

Q.E.D.

<u>REMARK</u>. If $p = 2$, L^2_k is a Hilbert space with respect to the inner product:

$$(f,g)_k = \sum_{0\leq|\alpha|\leq k} (D^{\alpha}f, D^{\alpha}g) = \sum_{0\leq|\alpha|\leq k} \int (D^{\alpha}f)\overline{(D^{\alpha}g)}\, dx$$

and $\|f\|_{2,k} = (f,f)^{1/2}_k$.

We have seen earlier that functions in L^p may have derivatives in L^p in the sense of distributions. These distribution derivatives are also called weak derivatives. We define now the so-called strong derivatives of functions in L^p and then we will show that, for $1 \leq p < \infty$, these two notions of derivative coincide.

Let $\{e_1,\ldots,e_n\}$ denote the standard basis in E^n, and h denote a real parameter.

DEFINITION. 2.2.3. Let $f \in L^p$, $1 \leq p < \infty$. If the differential quotient $(1/h)[f(x+he_i) - f(x)]$ converges in L^p norm, as $h \to 0$, to a function $g_i \in L^p$, then we say that $g_i = \frac{\partial f}{\partial x_i}$ is a strong L^p-derivative of f.

Strong L^p-derivatives of higher order are defined accordingly, by iteration. As usual, we write $\frac{\partial^2 f}{\partial x_j \partial x_i} = \frac{\partial}{\partial x_j}\left(\frac{\partial f}{\partial x_i}\right)$.

THEOREM. 2.2.4. Let $1 \leq p < \infty$. Then, L^p_k coincides with the class of functions $f \in L^p$ having strong L^p-derivatives $D^\alpha f$ for all α such that $|\alpha| \leq k$. Furthermore, a strong L^p-derivative $D^\alpha f$ coincides, as a distribution, with the corresponding derivative of f in the sense of distributions.

Proof. We shall treat only the case $k = 1$ since a repetition of the argument yields the result for any L^p_k. We begin by proving the last assertion of the theorem.

Let $f \in L_1^p$ and $g_i = \frac{\partial f}{\partial x_i}$ be a strong L^p-derivative of f. Viewing these functions as distributions and taking into account (as in the proof of 2.2.2) that convergence in L^p norm implies convergence in the sense of distributions, we have that, for all $u \in \mathcal{D}$,

$$\langle g_i, u\rangle = \lim_{h\to 0} \left\langle \frac{f(x+he_i) - f(x)}{h}, u \right\rangle =$$

$$= \lim_{h\to 0} \left\langle f, \frac{u(x-he_i) - u(x)}{h} \right\rangle =$$

$$= \left\langle f, -\frac{\partial u}{\partial x_i} \right\rangle$$

since, in the last step, we can take the limit inside the integral. Therefore, $g_i = \frac{\partial f}{\partial x_i}$ also in the sense of distributions.

Conversely, given $f \in L_1^p$ and $\frac{\partial f}{\partial x_i}$ a distribution derivative of f, we prove now that $\frac{\partial f}{\partial x_i}$ is also a strong L^p-derivative of f.

Since $\mathcal{D}$ is dense in L_1^p then, given any $\varepsilon > 0$, we can find a function $u \in \mathcal{D}$ such that $f = u + g$, where $\|g\|_{p,1} < \varepsilon$. Hence, by the triangle inequality,

$$\left\| \frac{f(x+he_i) - f(x)}{h} - \frac{\partial f}{\partial x_i} \right\|_p \leq \left\| \frac{u(x+he_i) - u(x)}{h} - \frac{\partial u}{\partial x_i} \right\|_p +$$

$$+ \left\| \frac{g(x+he_i) - g(x)}{h} - \frac{\partial g}{\partial x_i} \right\|_p = A + B.$$

Since $u \in \mathcal{D}$, the distribution derivative $\frac{\partial u}{\partial x_i}$ coincides with the ordinary partial derivative, so it is easy to see that, for some $\delta > 0$, $A < \varepsilon$ provided $|h| < \delta$.

For B, by the triangle inequality and the definition of L_1^p norm, we have the estimate

$$(1) \qquad B \leq \left\| \frac{g(x+he_i) - g(x)}{h} \right\|_p + \|g\|_{p,1} .$$

We now take a sequence $\{v_n\}$ in $\mathcal{D}$ such that $v_n \longrightarrow g$ in L_1^p norm, as $n \longrightarrow \infty$. Then, for every n,

$$\frac{v_n(x+he_i) - v_n(x)}{h} = \frac{1}{h}\int_0^h \frac{\partial v_n}{\partial x_i}(x+te_i)\, dt$$

so, using Minkowski's inequality for integrals, we obtain

$$\left\| \frac{v_n(x+he_i) - v_n(x)}{h} \right\|_p \leq \left\| \frac{\partial v_n}{\partial x_i} \right\|_p \leq \|v_n\|_{p,1} .$$

Consequently, letting $n \longrightarrow \infty$, we deduce that

$$(2) \qquad \left\| \frac{g(x+he_i) - g(x)}{h} \right\|_p \leq \|g\|_{p,1} .$$

Hence, from (1) and (2), we see, by definition of g, that

$$B \leq 2 \|g\|_{p,1} < 2\varepsilon$$

Combining our estimates, we conclude that for any $\varepsilon > 0$ there exists a $\delta > 0$ such that, for $|h| < \delta$,

$$\left\| \frac{f(x+he_i) - f(x)}{h} - \frac{\partial f}{\partial x_i} \right\|_p < 3\varepsilon .$$

Therefore, $\frac{\partial f}{\partial x_i}$ is also a strong L^p-derivative of f.

Q.E.D.

Using the preceding theorem we can obtain easily a useful general result. But first, let us recall some elementary notions.

For each vector $a \in E^n$, we define the translation operator τ_a by letting $[\tau_a f](x) = f(x-a)$. We say that an operator T <u>commutes with translations</u> if $T\tau_a = \tau_a T$, for every $a \in E^n$. For example, linear differential operators with constant coefficients,

$$P(D) = \sum_{|\alpha| \leq m} c_\alpha D^\alpha ,$$

on $C^\infty(E^n)$, commute with translations since differentiation and multiplication by a constant both do. Moreover, convolution operators commute with translations.

In fact, if $Tf = g*f = f*g$, then, for any $a \in E^n$,

$$[T(\tau_a f)](x) = [g*\tau_a f](x) = [\tau_a f*g](x) =$$

$$= \int f(y-a)\, g(x-y)\, dy = \quad \text{(letting } z = y-a\text{)}$$

$$= \int f(z)\, g(x-a-z)\, dz =$$

$$= (f*g)(x-a) = [\tau_a(Tf)](x) \quad .$$

This reasoning also shows that our singular convolution operators commute with translations. Hence, the following result will be of interest to us.

THEOREM. 2.2.5. Let $1 \leq p,q < \infty$ and suppose that T is a bounded linear operator from L^p to L^q which commutes with translations. Then, T commutes with differentiations. Furthermore, for all $k > 0$, T is bounded from L_k^p to L_k^q. In fact

$$\|Tf\|_{q,k} \leq \|T\| \, \|f\|_{p,k}$$

where $\|T\|$ is the norm of T as an operator from L^p to L^q.

Proof. For every $k > 0$, $L_k^p \subset L^p$, so T is defined on any L_k^p. As before, it suffices to prove the theorem in the case $k = 1$.

Given $f \in L_1^p$, we must show that

$$(1) \qquad \left[T\left(\frac{\partial f}{\partial x_i}\right)\right](x) = \left[\frac{\partial}{\partial x_i}(Tf)\right](x)$$

as elements of L^q , where we can view $\frac{\partial f}{\partial x_i}$ as a strong L^p-derivative and $\frac{\partial}{\partial x_i}(Tf)$ as a strong L^q-derivative.

Since T is linear and commutes with translations, we have that

$$(2) \quad T\left[\frac{f(y+he_i) - f(y)}{h}\right](x) = \frac{[Tf](x+he_i) - [Tf](x)}{h} \quad .$$

But, by continuity of T, the left-side of (2) converges, in L^q norm, to the left side of (1), as $h \rightarrow 0$. Hence, the right side of (2) converges in L^q norm, and, by definition of strong L^q derivative, it will converge to the right side of (1). Thus, T commutes with differentiations.

Consequently,

$$\|Tf\|_{q,1} = \left\{\sum_{|\alpha|\leq 1} \|D^\alpha(Tf)\|_q^2\right\}^{1/2} = \left\{\sum_{|\alpha|\leq 1} \|T(D^\alpha f)\|_q^2\right\}^{1/2} \leq$$

$$\leq \|T\| \left\{\sum_{|\alpha|\leq 1} \|D^\alpha f\|_p^2\right\}^{1/2} = \|T\| \, \|f\|_{p,1}$$

and the proof is complete.

Q.E.D.

Let us consider the Banach spaces $L_k^p = L_K^p(E^n)$, where $n \geq 2$ and $k \geq 0$ are integers. Our next objective will be to prove that for each p, $1 < p < \infty$, the spaces L_k^p are all isomorphic. For this purpose, we introduce a kind of integration operator J defined on the space $\mathcal{S}^*$ of all temperate distributions f by the formula

$$(Jf)^\wedge = d(\xi)^{-1}\,\hat{f}$$

where $d(\xi)$ is a strictly positive, infinitely differentiable radial function which coincides with $|\xi|$ for $|\xi| \geq 1$. In other words, J is the Fourier multiplier $J = F^{-1}d(\xi)^{-1}\,F$, where F and F^{-1} are respectively the Fourier and inverse Fourier transforms. We note that J has a 2-sided inverse, on $\mathcal{S}^*$, given by $J^{-1} = F^{-1}d(\xi)\,F$.

Let us recall that, for each $1 \leq j \leq n$, the j-th Riesz transform $R_j f$ (see [6], Chapter IV) is a singular convolution operator. Consequently, by Theorems 2.1.4 and 2.2.5, we have that, for $1 < p < \infty$ and all integers $k \geq 0$, R_j gives a bounded linear transformation of L_k^p into itself. Furthermore, if $f \in L^2$ for example, we have the formula

$$(R_j f)^\wedge = \frac{\xi_j}{|\xi|}\,\hat{f} \quad .$$

LEMMA. 2.2.6.

(a) For all finite $p \geq 1$ and all integers $k \geq 0$, $J: L^p_k \longrightarrow L^p_k$ is continuous.

(b) If $1 < p < \infty$, then $J: L^p_k \rightarrow L^p_{k+1}$ is continuous.

Proof. The proof depends first of all upon showing that $d(\xi)^{-1}$ is the Fourier transform of an integrable function.

Let $u_1(\xi)$ be a function in $\mathcal{D}$ which vanishes for $|\xi| \geq 1$ and such that $u_1(\xi) = 1$ near the origin. Then, the function $u_2(\xi) = d(\xi)^{-1} - |\xi|^{-1}\,[1-u_1(\xi)]$ coincides with $d(\xi)^{-1}$ near the origin and vanishes for $|\xi| \geq 1$. So, with u_1 and u_2 in $\mathcal{D}$, we can write

(1) $$d(\xi)^{-1} = |\xi|^{-1}\,[1-u_1(\xi)] + u_2(\xi) .$$

Since $n > 1$, we know that for some constant C the inverse Fourier transform of $|\xi|^{-1}$ is given by $C|x|^{1-n}$, which is a locally integrable function. If $j(x)$, v_1, v_2 denote the inverse Fourier transforms of $d(\xi)^{-1}$, u_1 and u_2 respectively, then (1) implies that

(2) $$j(x) = C|x|^{1-n} - C\,(|x|^{1-n} * v_1) + v_2(x)$$

where v_1 and v_2 belong to $\mathcal{S}$.

Moreover, the function $|x|^{1-n} * v_1$ is bounded, being the inverse Fourier transform of the integrable function $|\xi|^{-1} u_1(\xi)$. Consequently, (2) shows that $j(x)$ is locally integrable.

Since $d(\xi)^{-1} = F[j(x)]$, letting $\Delta = \sum_{k=1}^{n} \frac{\partial^2}{\partial \xi_k^2}$, we see that

$$(3) \qquad \Delta^m \, d(\xi)^{-1} = F[(2\pi i)^{2m} \, |x|^{2m} \, j(x)] \, .$$

But, for all integers m sufficiently large (certainly for all $m > \frac{n+1}{2}$), the left side of (3) is integrable, so, taking inverse Fourier transform, it follows that $|x|^{2m} \, j(x)$ is bounded for arbitrarily large m. Hence, $j(x)$ is a locally integrable function which is also rapidly decreasing at infinity. Therefore, $j(x)$ is integrable.

Now, we have that

$$Jf = F^{-1}[d(\xi)^{-1} \, F(f)] = j * f$$

where $j(x) \in L^1$, so Young's theorem implies that $J : L^p \longrightarrow L^p$ is a bounded operator for all p, $1 \leq p \leq \infty$. Furthermore, being a convolution operator, J commutes with translations, so Theorem 2.2.5 implies that $J : L^p_k \longrightarrow L^p_k$ is bounded for $1 \leq p < \infty$ and all positive integers k. Hence (a) is proved.

To prove (b) we observe that

$$F[D_j Jf](\xi) = \xi_j \, d(\xi)^{-1} \, \hat{f}(\xi)$$

so, using formula (1), we have

$$F[D_j Jf](\xi) = \frac{\xi_j}{|\xi|} \hat{f}(\xi) - \frac{\xi_j}{|\xi|} u_1(\xi) \, \hat{f}(\xi) + \xi_j u_2(\xi) \, \hat{f}(\xi) .$$

So, taking inverse Fourier transform, we obtain

$$(4) \qquad D_j Jf = R_j f - R_j K_1 f + K_2 f$$

where R_j is the j-th Riesz transform, $K_1 f = F^{-1}(u_1 \hat{f}) = v_1 * f$ is a convolution with kernel $v_1 = F^{-1}(u_1) \in \mathcal{S} \subset L^1$, and similarly, $K_2 f = F^{-1}(\xi_j u_2 \hat{f}) = w_2 * f$ is a convolution with the integrable kernel $w_2 = F^{-1}(\xi_j u_2) \in \mathcal{S}$. Consequently, for $i = 1,2$, Young's theorem and theorem 2.2.5 imply that $K_i : L_k^p \longrightarrow L_k^p$ is continuous for all $1 \leq p < \infty$ and all integers $k \geq 0$. In addition, we know that $R_j : L_k^p \longrightarrow L_k^p$ is continuous for all $1 < p < \infty$ and all $k \geq 0$. Therefore, we conclude from (4) that $D_j J : L_k^p \rightarrow L_k^p$ is continuous for all $1 < p < \infty$, integers $k \geq 0$, and $j = 1,\ldots,n$. Combining this conclusion with part (a), we obtain (b). Q.E.D.

<u>LEMMA</u>. 2.2.7. If $1 < p < \infty$, then $J^{-1} : L_{k+1}^p \longrightarrow L_k^p$ is continuous for every integer $k \geq 0$.

<u>Proof</u>. Let $f \in L_{k+1}^p$. Now, $F(J^{-1} f) = d(\xi) \hat{f}$ and, as in the preceding lemma we have that

$$(1) \qquad d(\xi) = |\xi|[1 - u_1(\xi)] + u_2(\xi)$$

where u_1 and u_2 belong to $\mathcal{D}$. Hence,

$$F(J^{-1}f) = |\xi|\,\hat{f} - |\xi|\,u_1(\xi)\,\hat{f} + u_2(\xi)\,\hat{f}$$

and, since $|\xi| = \sum_{j=1}^{n} \xi_j \frac{\xi_j}{|\xi|}$, we can write

$$F(J^{-1}f) = \sum_{j=1}^{n} \frac{\xi_j}{|\xi|}(\xi_j\hat{f}) - u_1(\xi)\sum_{j=1}^{n} \frac{\xi_j}{|\xi|}(\xi_j\hat{f}) + u_2(\xi)\,\hat{f} =$$

$$= g - u_1 g + u_2\hat{f} .$$

Consequently,

$$(2) \qquad J^{-1}f = F^{-1}(g) - F^{-1}(u_1 g) + F^{-1}(u_2\hat{f})$$

where we have set $g = \sum_{j=1}^{n} \frac{\xi_j}{|\xi|}(\xi_j\hat{f})$.

Now, $f \in L^p_{k+1}$ implies that $F^{-1}(\xi_j\hat{f}) = D_j f \in L^p_k$, so $F^{-1}(g) = \sum_{j=1}^{n} R_j(D_j f)$ is also in L^p_k and, from the continuity properties of the Riesz transforms R_j, we have that, for some constant $A_p > 0$ independent of f and k,

$$(3) \qquad \|F^{-1}(g)\|_{p,k} \leq A_p \sum_{j=1}^{n} \|D_j f\|_{p,k} \leq A_p\|f\|_{p,k+1} .$$

Letting v_1 and v_2 denote the inverse Fourier transforms of u_1 and u_2 respectively, we see that $F^{-1}(u_1 g) = v_1 * F^{-1}(g)$ and $F^{-1}(u_2 \hat{f}) = v_2 * f$, where v_1 and v_2 belong to $\mathcal{S} \subset L^1$. Hence, using Young's theorem, theorem 2.2.5 and estimate (3), we obtain

$$(4) \quad \|F^{-1}(u_1 g)\|_{p,k} \leq \|v_1\|_1 \; \|F^{-1}(g)\|_{p,k} \leq B_p \|f\|_{p,k+1} \; .$$

Finally,

$$(5) \quad \|F^{-1}(u_2 \hat{f})\|_{p,k} \leq \|v_2\|_1 \; \|f\|_{p,k} \leq C\|f\|_{p,k+1} \; .$$

Therefore, using estimates (3), (4) and (5) in formula (2), we deduce that $J^{-1}: L^p_{k+1} \longrightarrow L^p_k$ is continuous, for $1 < p < \infty$ and all integers $k \geq 0$.

Q.E.D.

Combining the two preceding lemmas, we obtain the following result:

<u>THEOREM</u>. 2.2.8. If $1 < p < \infty$, then the Banach spaces $L^p_k = L^p_k(E^n)$, $n \geq 2$, are isomorphic, for all integers $k \geq 0$.

<u>Proof</u>. The operator $J: L^p_k \longrightarrow L^p_{k+1}$ is a continuous linear isomorphism for all $k \geq 0$.

Q.E.D.

The preceding discussion has shown that, for $1 < p < \infty$, $L_1^p = J(L^p)$, and that, taking successive powers of J, $L_k^p = J(L_{k-1}^p) = \ldots = J^k(L^p)$ for every positive integer k. More generally, for any real number s, we can define on $\mathcal{S}^*$ the operator

$$J^s = F^{-1}\, d(\xi)^{-s}\, F .$$

These operators J^s for a commutative 1-parameter group, under composition, with $J^o = I$ being the identity operator. Furthermore, if $o < s < n$, the argument used to prove 2.2.6(a) shows that $J^s f = j_s * f$ is a convolution where the kernel $j_s = F^{-1}[d(\xi)^{-s}]$ is an integrable function. Consequently, for all $1 \leq p \leq \infty$, J^s is a bounded linear transformation of L^p into itself if $o \leq s < n$, and hence, by iteration, the conclusion holds for all real numbers $s \geq o$.

<u>DEFINITION</u>. 2.2.9. Let s be any real number and let $1 \leq p \leq \infty$. We define the class $L_s^p = L_s^p(E^n)$ to be the image of L^p under the operator J^s. We define the norm $\|f\|_{p,s}$ of any $f \in L_s^p$ in the following way. If $f \in L_s^p$ then $f = J^s g$ for some $g \in L^p$; in fact $g = J^{-s} f$ is unique and we set

$$\|f\|_{p,s} = \|g\|_p \quad .$$

<u>REMARK</u>. We note that if $s = k$ is a positive integer and if

$1 < p < \infty$, then L_s^p coincides with the Sobolev space L_k^p previously defined. If $|f|_{p,k}$ denotes the old norm in L_k^p , then, letting $g = J^{-k}f$ and using 2.2.6(b), we obtain

$$|f|_{p,k} = |J^k J^{-k} f|_{p,k} = |J^k g|_{p,k} \leq C\|g\|_p = C\|f\|_{p,k} .$$

Hence Banach's Open Mapping theorem, applied to the identity operator in L_k^p, implies that, conversely, $\|f\|_{p,k} \leq C_1|f|_{p,k}$ for some constant $C_1 > 0$. Therefore, the new norms are equivalent to the old ones.

<u>THEOREM</u>. 2.2.10.(a) The spaces L_s^p are Banach spaces isometrically isomorphic to L^p. (b) For all real numbers r and s, $J^r : L_s^p \longrightarrow L_{s+r}^p$ is an isometric isomorphism. (c) If $s \leq t$ then $L_s^p \supset L_t^p$ and the inclusion is continuous. (d) If $1 < p < \infty$, then D_j maps L_s^p continuously into L_{s-1}^p .

<u>Proof</u>. Part (a) is an immediate consequence of Definition 2.2.9. For (b) we note that since $L_s^p = J^s(L^p)$, then $J^r[L_s^p] = $
$= J^r[J^s(L^p)] = J^{s+r}(L^p) = L_{s+r}^p$.

To prove (c), let $f \in L_t^p$. Then $f = J^t g$ where $g \in L^p$ and $\|f\|_{p,t} = \|g\|_p$. Since $t - s \geq 0$ then, as we saw earlier, the operator J^{t-s} maps L^p continuously into itself. So, $J^{t-s}g$ is also in L^p and hence $f = J^s(J^{t-s}g)$ belongs to L_s^p .

Moreover,

$$\|f\|_{p,s} = \|J^{t-s}g\|_p \leq C\|g\|_p = C\|f\|_{p,t} .$$

Consequently, if $t \geq s$, L_t^p is continuously contained in L_s^p.

Finally, to prove (d), let us note that $D_j = F^{-1}\xi_j F$ commutes with all J^s. Thus, we can write

$$D_j = J^{s-1}\left(J^{1-s}D_j\right) = J^{s-1}(D_jJ)\ J^{-s} .$$

But, as we saw in the proof of 2.2.6(b), D_jJ maps L^p continuously into itself. Therefore, if $f = J^s g$ belongs to L_s^p, where $g \in L^p$, then $D_jf = J^{s-1}(D_jJ)g$ belongs to L_{s-1}^p and

$$\left\|D_jf\right\|_{p,s-1} = \left\|(D_jJ)g\right\|_p \leq C\|g\|_p = C\|f\|_{p,s} .$$

Q.E.D.

THEOREM. 2.2.11. Let $1 \leq p < \infty$ and let s be real.

(a) $\mathcal{S}$ is dense in every L_s^p.

(b) If $s < t$, then L_t^p is dense in L_s^p.

(c) $\mathcal{D}$ is dense in every L_s^p.

Proof. It is easy to see that, for any real s, J^s is a one-one map of $\mathcal{S}$ onto itself. Since $\mathcal{S}$ is dense in L^p and since $L_s^p = J^s(L^p)$ is a continuous image of L^p, conclusion (a) readily follows.

If $s < t$ then $L^p_t \subset L^p_s$, by theorem 2.2.10(c). Hence (b) follows directly from (a) and the inclusions $\mathcal{S} \subset L^p_t \subset L^p_s$.

By theorem 2.2.2, we know that $\mathcal{D}$ is dense in L^p_k for every integer $k \geq 0$ and any $1 \leq p < \infty$. Given any real s, choose a positive integer $k > s$. Then, in view of (b), the inclusions $\mathcal{D} \subset L^p_k \subset L^p_s$ are dense. Moreover, by 2.2.10(c), the inclusion $L^p_k \subset L^p_s$ is continuous, so, for some constant $C > 0$, $\|u\|_{p,s} \leq C\|u\|_{p,k}$. Now, given any $f \in L^p_s$ and any $\varepsilon > 0$, there exists $g \in L^p_k$ such that $\|f-g\|_{p,s} < \varepsilon$. Furthermore, we can choose $h \in \mathcal{D}$ such that $\|g-h\|_{p,k} < \varepsilon/C$. Hence,

$$\|f-h\|_{p,s} \leq \|f-g\|_{p,s} + \|g-h\|_{p,s} < \varepsilon + C\|g-h\|_{p,k} < 2\varepsilon .$$

Q.E.D.

In view of part (c) of theorem 2.2.10, we define L^p_∞ to be the intersection, over all real s, of L^p_s, $1 \leq p \leq \infty$. Equivalently, L^p_∞ may be defined as the intersection of the Sobolev spaces L^p_k for all positive integers k. By the preceding theorem, if $1 \leq p < \infty$, we see that $\mathcal{D} \subset L^p_\infty$ and hence that L^p_∞ is dense in L^p_s for any real s.

<u>THEOREM</u>. 2.2.12. (a) Let $f \in L^p_\infty$ and $g \in L^q_\infty$, where $1 \leq p < \infty$ and $q = p/(p-1)$. Consider the bilinear functional

$$\langle f,g \rangle = \int fg \, dx .$$

Then, for any real s, $| \langle f,g \rangle | \leq \|f\|_{p,s} \|g\|_{q,-s}$ and the functional $\langle f,g \rangle$ extends continuously to $L^p_s \times L^q_{-s}$.

(b) Every continuous linear functional on L^p_s is of the form $\lambda(f) = \langle f,g \rangle$ for some $g \in L^q_{-s}$.

Proof. We may assume that $s \geq 0$. If $0 < s < n$, then $J^s f = j_s * f$, where $j_s \in L^1$ is radial, being the inverse Fourier transform of the radial function $d(\xi)^{-s}$. Now if $f \in L^p_\infty$ and $g \in L^q_\infty$, then $J^s g$ is also in L^q_∞ and Hölder's inequality shows that $\langle f,J^s g \rangle$ exists. Thus, interchanging the order of integration and taking into account that j_s is radial, we see that

$$\langle f,J^s g \rangle = \int f(x) \left\{ \int j_s(x-y)\, g(y)\, dy \right\} dx =$$

$$= \int g(y) \left\{ \int j_s(x-y)\, f(x)\, dx \right\} dy =$$

$$= \int \left\{ \int j_s(y-x)\, f(x)\, dx \right\} g(y)\, dy =$$

$$= \langle J^s f,g \rangle .$$

Consequently, for all $s \geq 0$,

$$(1) \qquad \langle f,J^s g \rangle = \langle J^s f,g \rangle$$

since the case $s = 0$ is trivial and the case $s \geq n$ follows by iterating the case $0 < s < n$.

Since $f \in L^p_\infty$ implies that $J^{-s}f$ is also in L^p_∞, applying (1) with f replaced by $J^{-s}f$, we obtain

$$(2) \qquad \langle f,g \rangle = \langle J^s J^{-s} f,g \rangle = \langle J^{-s}f, J^s g \rangle .$$

Hence, by Hölder's inequality,

$$(3) \quad \begin{cases} |\langle f,g \rangle| = |\langle J^{-s}f, J^s g\rangle| \leq \\ \qquad \leq \|J^{-s}f\|_p \, \|J^s g\|_q = \|f\|_{p,s} \, \|g\|_{q,-s} . \end{cases}$$

Now, if $f \in L^p_s$ then $f = J^s(J^{-s}f)$ where $J^{-s} f \in L^p$, and if $g \in L^q_{-s}$ then $J^s g \in L^q$. Then, $\langle f,g \rangle$ extends by continuity to $L^p_s \times L^q_{-s}$ and satisfies (2), and hence (3). So, part (a) is proved.

To prove (b), let $\lambda(f)$ be a continuous linear functional on L^p_s. Since $f = J^s h$ where $h \in L^p$, it follows that $\lambda(f) = \lambda(J^s h)$ is a continuous linear functional on L^p, $1 \leq p < \infty$ and, hence, by a classical theorem of F. Riesz, it has the form $\lambda(J^s h) = \langle h,\tilde{g} \rangle$ where $\tilde{g} \in L^q$, $q = p/(p-1)$. Letting $g = J^{-s}\tilde{g}$, we see that $\tilde{g} = J^s g$ and $\lambda(f) = \lambda(J^s h) = \langle h,\tilde{g}\rangle = \langle h,J^s g\rangle = \langle J^s h,g\rangle = \langle f,g \rangle$, where $g \in L^q_{-s}$.

Q.E.D.

We remarked earlier that, if $1 < p < \infty$ and $s = k$ is a positive integer, then the spaces L^p_s coincide with the Sobolev spaces L^p_k of definition 2.2.1. We note also that, for $1 < p < \infty$ and any real s, $L^p_s = J^s(L^p)$ must be reflexive, being an isomorphic image of the reflexive space L^p. Moreover, theorem 2.2.12 shows that the dual of L^p_s can be identified with the space $L^{p'}_{-s}$, where now $p' = p/(p-1)$ denotes the conjugate exponent of p. In particular, L^p_{-k} can be identified with the dual of the Sobolev space $L^{p'}_k$. This result is sometimes taken as the definition of the spaces L^p_{-k} when $1 < p < \infty$ and k is a positive integer. Another useful characterization of these spaces L^p_{-k} is given by the following theorem.

__THEOREM__. 2.2.13. Let $1 < p < \infty$, and $k \geq 0$ be an integer. Then $g \in L^p_{-k}$ if and only if $g = \sum_{|\alpha| \leq k} D^\alpha g_\alpha$ for some $g_\alpha \in L^p$, where D^α are distribution derivatives.

__Proof__. By theorem 2.2.12 we can regard any $g \in L^p_{-k}$ as a bounded linear functional $g(f) = \langle f, g \rangle$, $f \in L^{p'}_k$. Consider the product space $\prod_{|\alpha| \leq k} L^{p'}$ with norm

$$|h| = \left\{ \sum_{|\alpha| \leq k} \|h_\alpha\|^2_{p'} \right\}^{1/2},$$

where $h = (h_\alpha)_{|\alpha| \leq k}$. Then the map $f \longrightarrow (D^\alpha f)_{|\alpha| \leq k}$ gives an isometric imbedding of $L^{p'}_k$ into $\prod_{|\alpha| \leq k} L^{p'}$, so, by the Hahn-Banach theorem, we can extend $g \in L^p_{-k}$ continuously to a bounded linear functional $\tilde{g}$ on $\prod_{|\alpha| \leq k} L^{p'}$. Since the dual of this product space is the product space $\prod_{|\alpha|\leq k} L^p$, we have that $\tilde{g} = (\tilde{g}_\alpha)_{|\alpha|\leq k}$, $\tilde{g}_\alpha \in L^p$.

Now, for all $f \in L_k^{p'}$,

$$\langle f,g \rangle = \langle f,\tilde{g} \rangle = \sum_{|\alpha|\leq k} \langle D^\alpha f,\tilde{g}_\alpha \rangle =$$

$$= \sum_{|\alpha|\leq k} \langle f,(-1)^{|\alpha|} D^\alpha \tilde{g}_\alpha \rangle =$$

$$= \langle f, \sum_{|\alpha|\leq k} D^\alpha (-1)^{|\alpha|} \tilde{g}_\alpha \rangle$$

so $g = \sum_{|\alpha|\leq k} D^\alpha g_\alpha$, where $g_\alpha = (-1)^{|\alpha|} \tilde{g}_\alpha \in L^p$.

Conversely, any g of this form belongs to L_{-k}^p, by parts (c) and (d) of theorem 2.2.10.

Q.E.D.

DEFINITION. 2.2.14. Let $1 < p,q < \infty$ and s and t be arbitrary real numbers. Given a bounded linear operator $T:L_s^p \longrightarrow L_t^q$ we define the transpose $T':L_{-t}^{q'} \longrightarrow L_{-s}^{p'}$ to be the induced bounded linear operator on the dual spaces given by

$$\langle Tf,g \rangle = \langle f,T'g \rangle$$

for all $f \in L_s^p$ and $g \in L_{-t}^{q'}$.

As is well known, T and T' have the same norm.

If $s \geq 0$ and $a \in E^n$, we recall that the translation operator τ_a is defined on L_s^p by $[\tau_a f](x) = f(x-a)$.

If $f \in L^p_{-s}$, we define $\tau_a f$ to be the element of L^p_{-s} given by

$$\langle \tau_a f, g \rangle = \langle f, \tau_{-a} g \rangle$$

for all $g \in L^{p'}_s$. It is easy to see that both definitions coincide when $s = 0$. Similarly, if $f \in L^p_s$, $s \geq 0$, we define the operator ρ by $[\rho f](x) = f(-x)$. If $f \in L^p_{-s}$, we define ρf to be the element of L^p_{-s} given by

$$\langle \rho f, g \rangle = \langle f, \rho g \rangle$$

for all $g \in L^{p'}_s$.

The differentiation operator D_j (in the sense of distributions) is defined on all spaces L^p_s since, as we know, $L^p_s \subset \mathcal{S}^*$. Also, by theorem 2.2.10(d), if $1 < p < \infty$ then $D_j : L^p_s \to L^p_{s-1}$ is a bounded linear operator. For all f and g in $\mathcal{D}$, integration by parts shows that

$$\langle D_j f, g \rangle = \langle f, -D_j g \rangle .$$

More generally, since for $1 < p < \infty$ $\mathcal{D}$ is dense in all L^p_s, the preceding formula is also valid, by continuity, for all $f \in L^p_s$ and all $g \in L^{p'}_{1-s}$. Consequently, the transpose of D_j is given by $-D_j$, and, in general, the transpose of D^α is $(-1)^{|\alpha|} D^\alpha$.

Let us recall that if $1 \leq p \leq \infty$ then, for all $s \geq 0$, $J^s = F^{-1} d(\xi)^{-s} F$ is a continuous linear map of L^p into itself, which is also well-defined on $\mathcal{S}^*$.

If $1 < p < \infty$, any singular convolution operator K gives a continuous linear map of L^p into itself, and furthermore, on $\mathcal{S}, K:\mathcal{S} \to \mathcal{S}^*$ can be expressed in the form $K = F^{-1}h(\xi) F$ where, if k denotes the kernel of K, $h = F(p.v.k)$ is a bounded function (cf. theorems 2.1.4 and 2.1.6).

THEOREM. 2.2.15. Let $1 < p < \infty$ and $s \geq 0$. If K is a singular convolution operator, then

(a) $KJ^s = J^sK$ on L^p.

(b) $K:L^p_s \to L^p_s$ is a bounded linear operator with norm $\|K\|_s \leq \|K\|$, where $\|K\|$ is the norm of K as an operator on L^p.

Proof. Let $h = F(p.v.k)$ where k is the kernel of K. Since $J^s:\mathcal{S} \to \mathcal{S}$ and $K = F^{-1}h(\xi) F$ on $\mathcal{S}$, we have that for all $f \in \mathcal{S}$

$$KJ^sf = KF^{-1}d(\xi)^{-s}\hat{f} = F^{-1}h(\xi)\, d(\xi)^{-s}\hat{f} =$$

$$= (F^{-1}d(\xi)^{-s} F)(F^{-1}h(\xi) F)\, f = J^sKf .$$

Since $\mathcal{S}$ is dense in L^p and both J^s and K are continuous maps of L^p into itself, it follows that $KJ^sf = J^sKf$, for all $f \in L^p$. So (a) is proved.

Since $s \geq 0$, $L^p_s \subset L^p$ so K is defined on L^p_s. If $f \in L^p_s$, then $f = J^sg$, where $g \in L^p$ and $\|g\|_p = \|f\|_{p,s}$. Hence, using (a), we obtain

$$\|Kf\|_{p,s} = \|KJ^s g\|_{p,s} = \|J^s Kg\|_{p,s} =$$

$$= \|Kg\|_p \le \|K\| \, \|g\|_p = \|K\| \, \|f\|_{p,s}$$

which proves (b).

Q.E.D.

<u>LEMMA</u>. 2.2.16. Let $1 < p < \infty$ and $K:L^p \to L^p$ be a singular convolution operator with kernel $k(x)$. Then the transpose $K':L^{p'} \to L^{p'}$ is a singular convolution operator with kernel $[\rho k](x) = k(-x)$. In particular, $K' = K$ if $k(x)$ is an even function and $K' = -K$ if $k(x)$ is odd.

<u>Proof</u>. By definition, $\langle f,K'g\rangle = \langle Kf,g\rangle$ for all $f \in L^p$ and $g \in L^{p'}$. By continuity of K and K' it suffices to consider f and g in $\mathcal{D}$. Then, if $k_\varepsilon(x) = k(x)$ for $\varepsilon < |x| < \frac{1}{\varepsilon}$ $k_\varepsilon(x) = 0$ otherwise, we obtain

$$\begin{aligned}
\langle Kf,g\rangle &= \lim_{\varepsilon\to 0} \langle k_\varepsilon * f,g\rangle = \\
&= \lim_{\varepsilon\to 0} \int \left\{ \int k_\varepsilon(x-y)\, f(y)\, dy\right\} g(x)\, dx = \\
&= \lim_{\varepsilon\to 0} \int f(y) \left\{ \int k_\varepsilon(x-y)\, g(x)\, dx\right\} dy = \\
&= \lim_{\varepsilon\to 0} \langle f,\rho k_\varepsilon * g\rangle
\end{aligned}$$

and the conclusion follows.

Q.E.D.

<u>THEOREM</u>. 2.2.17. Let $1 < p < \infty$ and $s > 0$. Any singular convolution operator $K: L^p \longrightarrow L^p$, with norm $\|K\|$, can be extended uniquely to a continuous linear map of L^p_{-s} into itself with norm $\leq \|K\|$.

<u>Proof</u>. Since $K: L^p \longrightarrow L^p$ is bounded, so is its transpose $K': L^{p'} \longrightarrow L^{p'}$ and $\|K'\| = \|K\|$. By the preceding lemma, K' is also a singular convolution operator, so theorem 2.2.15 implies that $K': L^{p'}_s \longrightarrow L^{p'}_s$ is bounded with norm $\|K'\|_s \leq \|K'\|$.

Let $\tilde{K}$ be the transpose of the restriction of K' to $L^{p'}_s$. Then, $\tilde{K}: L^p_{-s} \longrightarrow L^p_{-s}$ is bounded with norm $\|\tilde{K}\| = \|K'\|_s \leq \|K'\| = \|K\|$. Moreover, for all $f \in L^p$ and $g \in L^{p'}_s$ which is dense in $L^{p'}$, we have, by definition of $\tilde{K}$, that

$$\langle \tilde{K}f, g \rangle = \langle f, K'g \rangle = \langle Kf, g \rangle .$$

Consequently, $\tilde{K} = K$ on L^p.

Since L^p is dense in L^p_{-s} and $\tilde{K}$ is continuous on L^p_{-s}, it follows that this extension is unique.

Q.E.D.

The following result, which supplements theorem 2.2.5, is an analogue of the preceding theorem.

<u>THEOREM</u>. 2.2.18. Let $1 < p,q < \infty$ and k be a positive integer. If $T:L^p \longrightarrow L^q$ is a bounded linear operator which commutes with translations, then T has a unique bounded extension $\tilde{T}:L^p_{-k} \longrightarrow L^q_{-k}$ which commutes with translations and differentiations. Moreover, $\|\tilde{T}\| \leq \|T\|$.

<u>Proof</u>. The transpose $T':L^{q'} \longrightarrow L^{p'}$ is bounded with $\|T'\| = \|T\|$ and also commutes with translations. In fact, for all $f \in L^p$ and $g \in L^{q'}$.

$$\langle f,T'\tau_a g\rangle = \langle \tau_{-a}Tf,g\rangle =$$

$$= \langle T\tau_{-a}f,g\rangle = \langle f,\tau_a T'g\rangle$$

since T commutes with translations. Consequently, by theorem 2.2.5, $T':L^{q'}_k \longrightarrow L^{p'}_k$ is bounded with norm $\|T'\|_k \leq \|T'\|$ and commutes with differentiations.

Let $\tilde{T}$ be the transpose of the restriction of T' to $L^{q'}_k$. Then $\tilde{T}:L^p_{-k} \longrightarrow L^q_{-k}$ is bounded with norm $\|\tilde{T}\| = \|T'\|_k \leq \|T'\| = \|T\|$ and, by the previous argument, $\tilde{T}$ commutes with translations. Let us verify that $\tilde{T}$ commutes with differentiations.

For all $f \in L^p_{1-k}$ and $g \in L^{q'}_k$ which is dense in $L^{q'}_{k-1}$,

$$\langle \tilde{T}D_j f,g \rangle = \langle D_j f,T'g \rangle = -\langle f,D_j T'g \rangle =$$

$$= -\langle f,T'D_j g \rangle = \langle D_j \tilde{T}f,g \rangle$$

since T' commutes with differentiations. The rest of the proof now follows as in the preceding theorem.

Q.E.D.

Let us agree to call a bounded linear operator $T:L^p \longrightarrow L^q$ translation invariant if T commutes with translations. The next two theorems, due to Hörmander (cf. [4]), show that such operators which are non-trivial exist only for $p \leq q$, and secondly that they are essentially convolutions with a kernel which is a temperate distribution.

THEOREM. 2.2.19. Let $1 \leq p,q < \infty$ and $T:L^p \longrightarrow L^q$ be a translation invariant bounded linear operator. If $p > q$ then $T = 0$.

Proof. We note first that if $u \in L^r$, $1 \leq r < \infty$, then

$$(1) \qquad \|u+\tau_a u\|_r \longrightarrow 2^{1/r}\|u\|_r \ , \text{ as } |a| \longrightarrow \infty .$$

This can be seen as follows. Given any $\varepsilon > 0$ we can write $u = v + w$, where v has compact support and $\|w\|_r < \varepsilon$. Since for all $|a|$ large enough the supports of v and of $\tau_a v$ are disjoint, we have that

(2) $$\|v+\tau_a v\|_r = 2^{1/r} \|v\|_r \quad , \text{ if } |a| \text{ is large.}$$

Now, since $w + \tau_a w = (u+\tau_a u) - (v+\tau_a v)$ and $\|\tau_a w\|_r = \|w\|_r < \varepsilon$, it follows that

$$\Big| \|u+\tau_a u\|_r - \|v+\tau_a v\|_r \Big| \leq \|w+\tau_a w\|_r \leq 2\|w\|_r < 2\varepsilon$$

so, on account of (2), we see that

$$\Big| \|u+\tau_a u\|_r - 2^{1/r}\|v\|_r \Big| < 2\varepsilon \ , \text{ if } |a| \text{ is large.}$$

Consequently, since $\Big| \|v\|_r - \|u\|_r \Big| \leq \|w\|_r < \varepsilon$,

$$\Big| \|u+\tau_a u\|_r - 2^{1/r}\|u\|_r \Big| \leq \Big| \|u+\tau_a u\|_r - 2^{1/r}\|v\|_r \Big| + 2^{1/r} \varepsilon \leq 4\varepsilon$$

for $|a|$ large, and (1) is proved.

Clearly, $C = \|T\|$ is the smallest number such that

(3) $$\|Tf\|_q \leq C\|f\|_p \ , \text{ for all } f \in L^p .$$

Moreover, since T is linear and translation invariant,

$$\|Tf+\tau_a Tf\|_q = \|T(f+\tau_a f)\|_q \leq C\|f+\tau_a f\|_p$$

hence, letting $|a| \longrightarrow \infty$ and using (1), we obtain

(4) $$\|Tf\|_q \leq 2^{(1/p)-(1/q)} C\|f\|_p .$$

But, if $1 \leq q < p < \infty$ then $\frac{1}{p} - \frac{1}{q} < 0$ so, if $C > 0$, the constant in (4) is less than C contradicting (3). Therefore, if $p > q$, we must have that $C = \|T\| = 0$, that is $T = 0$.

Q.E.D.

Next, we need a lemma which is a very special case of some classical results of Sobolev.

<u>LEMMA</u>. 2.2.20. If a function v on E^n and its derivatives of order $\leq n$ are locally in L^p, $1 \leq p \leq \infty$, then, after correction on a set of measure zero, v is continuous and, for some constant $C > 0$,

$$|v(x)| \leq C \sum_{|\alpha| \leq n} \left\{ \int_{|y-x| \leq 1} |D^\alpha v|^p \, dy \right\}^{1/p}$$

<u>Proof</u>. Since, by Hölder's inequality, functions which are locally in L^p are locally in L^1 and, for any compact set K,

$$\int_K |D^\alpha v| \, dy \leq |K|^{\frac{p-1}{p}} \left\{ \int_K |D^\alpha v|^p \, dy \right\}^{1/p}$$

it suffices to prove the lemma for $p = 1$.

Let Q^n denote the set of all points in E^n with rational coordinates. Fixing a point $x \in Q^n$, we let $w = uv$ where $u(y)$ is a function in $\mathcal{D}$ which equals 1 on a neighborhood of x and vanishes if $|y-x| > 1$. If $H(t)$ is the Heaviside function, that is $H(t) = 1$ if $t > 0$ and $H(t) = 0$ if $t < 0$, we let $h(y) = H(y_1)\ldots H(y_n)$. To simplify notation, let us write $\partial^n = \frac{\partial}{\partial y_n} \cdots \frac{\partial}{\partial y_1}$. Then, in the sense of distributions, $\partial^n h = \delta$, the Dirac measure, and

$$w = w * \delta = w * \partial^n h = \partial^n w * h .$$

Since w has compact support it follows, by hypothesis, that $\partial^n w$ is integrable. Hence, since h is bounded, Lebesgue's Dominated Convergence theorem implies that the convolution $\partial^n w * h$ is a continuous function. Consequently, as a function, w coincides with $\partial^n w * h$ almost everywhere so, correcting it on a set of measure zero, we can make w continuous. Therefore, on a neighborhood of x, $v = w$ is continuous. Letting x vary in Q^n, we deduce that, after correction on a set of measure zero, v is continuous everywhere.

Moreover,

$$v(x) = w(x) = \int_{|y-x| \leq 1} \partial^n w(y)\, h(x-y)\, dy$$

so, since $|h(y)| \leq 1$, we obtain, using Leibnitz's formula,

$$|v(x)| \leq \int_{|y-x|\leq 1} |\partial^n w| \, dy \leq C \sum_{|\alpha|\leq n} \int_{|y-x|\leq 1} |D^\alpha v| \, dy$$

which is the desired estimate.

Q.E.D.

THEOREM. 2.2.21. Let $1 \leq p,q < \infty$ and $T: L^p \to L^q$ be a translation invariant bounded linear operator. Then there exists a unique $f \in L^{p'}_{-n}$ such that, for all $u \in \mathcal{S}$, $Tu = f * u$.

Proof. By theorem 2.2.5, we know that T commutes with differentiations and is a bounded operator from L^p_n to L^q_n. If $u \in L^p_n$ then $Tu \in L^q_n$ so, by the preceding lemma, we deduce, after correction on a set of measure zero, that Tu is a continuous function and that

$$|[Tu](0)| \leq C \sum_{|\alpha|\leq n} \|D^\alpha Tu\|_q \leq C_1 \sum_{|\alpha|\leq n} \|D^\alpha u\|_p .$$

This shows that $[Tu](0)$ is a continuous linear functional on L^p_n, hence, by theorem 2.2.12, there exists a unique element $g \in L^{p'}_{-n}$ such that $[Tu](0) = \langle u,g \rangle$.

Letting $g = \rho f$, we see that in particular, for all $u \in \mathcal{S}$, $\langle u,g \rangle = \langle u,\rho f \rangle = \langle \rho u,f \rangle = \int f(y)\, u(-y)\, dy$. Therefore,

$$[Tu](0) = (f * u)(0)$$

for all $u \in \mathcal{S}$. In view of the translation invariance of both sides of this formula, we conclude that $[Tu](x) = (f * u)(x)$ for every x.

Q.E.D.

Finally, the following result follows directly from lemma 2.2.20.

<u>COROLLARY</u>. 2.2.22. Let $1 \leq p \leq \infty$. Then $f \in L^p_\infty$ if and only if f coincides almost everywhere with a C^∞ function such that it and all its derivatives belong to L^p.

EXERCISES

Let $1 < p < \infty$, R_j denote the j-th Riesz transform, and $D_j = (2\pi i)^{-1} \frac{\partial}{\partial x_j}$, $j = 1,2,\ldots,n$.

1. Prove that: (a) $R_j R_k = R_k R_j$ on L^p

(b) $R_j D_k = D_k R_j$ on L_1^p

(c) $R_j D_k = R_k D_j$ on L_1^p.

2. Define the operator Λ by the formula

$$\Lambda = \sum_{j=1}^{n} R_j D_j .$$

Prove that: (a) For all $s \geq 0$, $\Lambda : L_{s+1}^p \to L_s^p$ is continuous; $\Lambda R_j = R_j \Lambda$ on L_1^p, and $\Lambda D_j = D_j \Lambda$ on L_2^p.

(b) $R_j \Lambda = D_j$ on L_1^p

(c) $(2\pi i \Lambda)^2 = \Delta$ on L_2^p, where $\Delta = \sum_{j=1}^{n} \frac{\partial^2}{\partial x_j^2}$ is the Laplacian.

3. Prove that: (a) $\Lambda = F^{-1} |\xi| F$ on L_1^2.

(b) If I denotes the identity operator on L^p, then $\Lambda J = I + S$ where $S : L^p \to L_\infty^p$ is continuous. [Hint: use formula (4) of lemma 2.2.6].

CHAPTER III

SPHERICAL HARMONICS

Spherical harmonics play a useful role in many problems of classical analysis. We introduce them at this stage in order to be able to exploit their properties in proving the fundamental result of the next chapter (Theorem 4.1.5). The presentation which follows is a slightly expanded version of the elegant discussion of this topic given by Calderón in [1].

Let us denote by Π_m the set of all polynomials in $x = (x_1, \ldots, x_n)$ which are homogeneous of degree m. Here, $n \geq 2$. For simplicity, we shall assume throughout the chapter that our polynomials have real coefficients (the complex case being of course identical except for the occasional presence of complex conjugates).

LEMMA. 3.0. Π_m is a finite dimensional real vector space. If we denote by a $g(m)$ the dimension of Π_m, then

$$g(m) = \binom{m + n - 1}{n - 1}.$$

Proof. The first statement is obvious. Clearly, the monomials x^α, with $|\alpha| = m$, form a basis in Π_m, so the dimension $g(m)$ is equal to the number of such monomials. Since these monomials occur as coefficients in

$$\prod_{i=1}^{n} (1 - x_i t)^{-1} = \prod_{i=1}^{n} (1 + x_i t + \cdot \cdot \cdot + x_i^k t^k + \cdot \cdot \cdot)$$

their number is equal to the coefficient of t^m in $(1 - t)^{-n}$. Therefore,

$$g(m) = \frac{1}{m!} \frac{d^m}{dt^m} (1 - t)^{-n}\Big|_{t = 0} =$$

$$= \frac{1}{m!} n(n + 1) \ . \ . \ . \ (n + m - 1)(1 - t)^{-(n + m)}\Big|_{t = 0} =$$

$$= \frac{(n + m - 1)!}{m!(n - 1)!} = \binom{n + m - 1}{n - 1} .$$

Q.E.D.

With each polynomial $P(x) = \sum_{|\alpha|=m} a_\alpha x^\alpha$ in Π_m we can associate the (homogeneous) differential polynomial $P\left(\frac{\partial}{\partial x}\right) = \sum_{|\alpha|=m} a_\alpha \left(\frac{\partial}{\partial x}\right)^\alpha$ obtained from $P(x)$ by replacing each monomial x^α by the corresponding differential monomial $\left(\frac{\partial}{\partial x}\right)^\alpha$. Let us note that for $|\alpha| = |\beta|$

$$\left(\frac{\partial}{\partial x}\right)^\alpha x^\beta = \begin{cases} \alpha! & \text{if } \alpha = \beta \\ 0 & \text{otherwise.} \end{cases}$$

<u>LEMMA</u>. 3.1. Every Π_m is a (real) inner product space.

<u>Proof</u>. From the preceding remark it is clear that, for all P

and Q in Π_m,

$$\langle P,Q\rangle = P\left(\frac{\partial}{\partial x}\right) Q(x)$$

is a bilinear form on $\Pi_m \times \Pi_m$. Also, if $P(x) = \sum a_\alpha x^\alpha$ and $Q(x) = \sum b_\beta x^\beta$, $|\alpha| = |\beta| = m$, then

$$\langle P,Q\rangle = \sum a_\alpha b_\alpha \alpha! = \sum b_\alpha a_\alpha \alpha! = \langle Q,P\rangle$$

$$\langle P,P\rangle = \sum a_\alpha^2 \alpha! \geq 0$$

$$\langle P,P\rangle = 0 \text{ if and only if } P(x) \equiv 0.$$

Q.E.D.

Let $P \in \Pi_\ell$ and $Q \in \Pi_m$, $\ell \leq m$. Then $\langle P,Q\rangle = P\left(\frac{\partial}{\partial x}\right) Q(x)$ belongs to $\Pi_{m-\ell}$. Hence for each $P \in \Pi_\ell$ the associated differential polynomial $P\left(\frac{\partial}{\partial x}\right)$ defines a linear operator $P\left(\frac{\partial}{\partial x}\right): Q(x) \rightarrow P\left(\frac{\partial}{\partial x}\right) Q(x)$ which maps Π_m into $\Pi_{m-\ell}$.

<u>LEMMA</u>. 3.2. Let $\ell \leq m$, $P \in \Pi_\ell$, $P(x) \not\equiv 0$. Then the linear map $P\left(\frac{\partial}{\partial x}\right): \Pi_m \rightarrow \Pi_{m-\ell}$ is onto.

<u>Proof</u>. Suppose on the other hand that the range $P\left(\frac{\partial}{\partial x}\right) \Pi_m$ is a proper subspace V of $\Pi_{m-\ell}$. Then, by Lemma 3.1, there exists an element $R(x) \not\equiv 0$ in $\Pi_{m-\ell}$ which is orthogonal to V. In other words, for every $Q \in \Pi_m$,

$$\left\langle R, P\left(\frac{\partial}{\partial x}\right) Q(x)\right\rangle = R\left(\frac{\partial}{\partial x}\right) P\left(\frac{\partial}{\partial x}\right) Q(x) \equiv 0.$$

Choosing $Q(x) = R(x)\,P(x)$ above, we see that $\langle Q,Q \rangle = 0$ and hence $Q(x) = R(x)\,P(x) \equiv 0$. Since $P(x) \not\equiv 0$ by hypothesis, this implies that $R(x) \equiv 0$, which is a contradiction.

Q.E.D.

<u>THEOREM</u>. 3.3. (Decomposition Theorem). Let $\ell \leq m$, $P \in \Pi_\ell$, $P(x) \not\equiv 0$. Then, every $T \in \Pi_m$ can be decomposed uniquely in the form

$$T(x) = \sum_k P^k(x)\, R_k(x)$$

where $R_k \in \Pi_{m-k\ell}$, $P\left(\frac{\partial}{\partial x}\right) R_k(x) \equiv 0$, and we sum over all integers $k \geq 0$ such that $k\ell \leq m$. Moreover the $R_k(x) \not\equiv 0$ are not divisible by P.

<u>Proof</u>. Let us show first that T can be decomposed uniquely in the form $T = PS_1 + R_o$ where $S_1 \in \Pi_{m-\ell}$ and $P\left(\frac{\partial}{\partial x}\right) R_o(x) \equiv 0$.

Consider the subspace

$$M = \{PS : S \in \Pi_{m-\ell}\}$$

of all elements of Π_m which are divisible by P, and the subspace

$$N = \{R \in \Pi_m : P\left(\frac{\partial}{\partial x}\right) R(x) \equiv 0\}$$

of all elements of Π_m which are annihilated by $P\left(\frac{\partial}{\partial x}\right)$. Now, $M \cap N = \{0\}$ since, if $PS \in N$, then $P\left(\frac{\partial}{\partial x}\right) P(x) S(x) \equiv 0$ so that $S\left(\frac{\partial}{\partial x}\right) P\left(\frac{\partial}{\partial x}\right) P(x) S(x) = \langle PS,PS \rangle = 0$, which implies that $PS = 0$. Hence, $\dim(M \oplus N) = \dim(M) + \dim(N) = g(m - \ell) + \dim(N)$. But, from Lemma 3.2, it follows that $g(m - \ell) + \dim(N) = \dim(\Pi_m)$, so $\dim(M \oplus N) = \dim(\Pi_m)$. Consequently, $\Pi_m = M \oplus N$ and therefore each $T \in \Pi_m$ can be decomposed uniquely in the form $T = PS_1 + R_o$ with $S_1 \in \Pi_{m-\ell}$ and $R_o \in N$. Since $M \cap N = \{0\}$, R_o is not divisible by P, unless $R_o(x) \equiv 0$.

Repeating on S_1 the argument used for T, we obtain the unique decomposition $S_1 = PS_2 + R_1$, with $S_2 \in \Pi_{m-2\ell}$ and R_1 annihilated by $P\left(\frac{\partial}{\partial x}\right)$ and not divisible by P (unless $R_1 = 0$). This recursive process ends when we reach an S_k of degree $<\ell$, thus S_k is not divisible by P and $P\left(\frac{\partial}{\partial x}\right) S_k(x) \equiv 0$, so we may set $S_k = R_k$.

Q.E.D.

Of particular interest to us is the homogeneous polynomial $|x|^2 = x_1^2 + x_2^2 + \cdots + x_n^2$, of degree 2, whose corresponding differential operator is the Laplacian $\Delta = \sum_{j=1}^{n}\left(\frac{\partial^2}{\partial x_i^2}\right)$.

<u>DEFINITION</u>. 3.4. We shall call <u>solid harmonics of degree m</u> all $P \in \Pi_m$ which satisfy Laplace's equation $\Delta P = 0$. The restrictions of these solid harmonics to the unit sphere will be called <u>spherical harmonics of degree m</u>.

We shall use the following notation:

$\{S_m\}$ = all solid harmonics of degree m.

$\{Q_m\}$ = all spherical harmonics of degree m.

REMARKS. 3.5.

a) From Definition 3.4 and Lemma 3.2 we see that $\{S_m\}$ is the nullspace of the Laplacian as a linear map from Π_m onto Π_{m-2}. If $m = 0, 1$ then $\{S_m\} = \Pi_m$. For $m \geq 2$, $\{S_m\}$ is a proper subspace of Π_m of dimension

$$d(m) = g(m) - g(m-2).$$

Clearly this last expression is valid for all integers $m \geq 0$ provided that we set $g(-1) = g(-2) = 0$.

b) If $S_m \in \{S_m\}$ then, for all $x \neq 0$,

$$S_m(x) = S_m\left(|x| \frac{x}{|x|}\right) = |x|^m S_m\left(\frac{x}{|x|}\right) = |x|^m Q_m(x')$$

where $Q_m(x') = S_m(x')$ is the corresponding spherical harmonic in $\{Q_m\}$. Since this correspondence is an isomorphism, it follows from a) that $\{Q_m\}$ is a vector space of dimension $d(m) = g(m) - g(m-2)$.

c) On E^2, $g(m) = \begin{pmatrix} m+1 \\ 1 \end{pmatrix} = m + 1$, $g(m-2) = m - 1$ if $m > 0$, so $d(m) = g(m) = m + 1$ if $m = 0, 1$, whereas for $m \geq 2$, $d(m) = g(m) - g(m-2) = 2$. In terms of the complex variable

$z = x + iy = re^{i\theta}$, we have that $z^m = r^m \cos m\theta + ir^m \sin m\theta$. Therefore, for every m, $\cos m\theta$ and $\sin m\theta$ are the only linearly independent spherical harmonics on the plane. On E^3, it is easy to see that $d(m) = 2m + 1$, so the dimension of $\{S_m\}$, and hence of $\{Q_m\}$, increases as the odd integers.

d) In general, on E^n, we have that

$$g(m) = \binom{m + n - 1}{n - 1} = \frac{(m + n - 1)(m + n - 2)\cdot\ \cdot\ \cdot(m + 1)}{(n - 1)!} \sim \frac{m^{n-1}}{(n - 1)!}$$

as $m \to \infty$. Using the Mean Value Theorem, we obtain

$$d(m) = g(m) - g(m - 2) \sim \left[\frac{m^{n-1}}{(n - 1)!} - \frac{(m - 2)^{n-1}}{(n - 1)!}\right] \sim 2\, \frac{m^{n-2}}{(n - 2)!}\ .$$

Therefore, for any $n \geq 2$, we conclude that

$$d(m) \sim Cm^{n-2}$$

as $m \to \infty$, where the constant C depends only on n.

<u>COROLLARY</u>. 3.6. Any continuous function on Σ can be approximated uniformly by a finite linear combination of spherical harmonics.

<u>Proof</u>. Let f be continuous on Σ. We may assume that f is

real valued. According to Weierstrass Approximation Theorem, f can be approximated uniformly by the restriction to Σ of a polynomial T. Since T is a finite sum of homogeneous pieces, it suffices to consider each piece $T_m \in \Pi_m$. For $m = 0$ and $m = 1$ the conclusion is obvious. If $m \geq 2$ then, using the Decomposition Theorem with $P(x) = |x|^2$, we obtain the finite sum

$$T_m(x) = \sum |x|^{2k} R_k(x), \qquad 2k \leq m$$

where R_k are solid harmonics of degree $m - 2k$. Hence, on the unit sphere, $T_m(x') = \sum_{2k\leq m} R_k(x')$, where $R_k(x')$ belong to $\{Q_{m-2k}\}$.

Q.E.D.

<u>LEMMA</u>. 3.7. Let $Q_i \in \{Q_i\}$ and $S_i(x) = |x|^i Q_i(x')$ be the corresponding solid harmonics. Then, for $k \neq m$,

a) $\int_\Sigma Q_k Q_m \, dx' = 0$, b) $\int_{|x|\leq 1} S_k S_m \, dx = 0$.

<u>Proof</u>. Since, in polar coordinates,

$$\int_{|x|\leq 1} S_k S_m \, dx = \int_0^1 r^{k+m} r^{n-1} \, dr \int_\Sigma Q_k Q_m \, dx'$$

it suffices to prove a).

According to Green's formula,

$$\int_{|x|\leq 1} \{S_k \Delta S_m - S_m \Delta S_k\}\, dx = \int_\Sigma \{S_k \frac{\partial}{\partial\nu} S_m - S_m \frac{\partial}{\partial\nu} S_k\}\, dx'$$

where ν is the outer normal to Σ, so $\frac{\partial}{\partial\nu} = \frac{\partial}{\partial|x|}$ is the radial derivative. Hence,

$$\frac{\partial}{\partial\nu} S_i(x) = i|x|^{i-1} Q_i(x').$$

Moreover, by definition of solid harmonics, $\Delta S_i = 0$. Therefore,

$$0 = \int_\Sigma \{Q_k m Q_m - Q_m k Q_k\}\, dx' = (m - k) \int_\Sigma Q_k Q_m\, dx'.$$

Q.E.D.

Let us regard the vector spaces $\{Q_m\}$ as linear subspaces of the real Hilbert space $L^2(\Sigma)$, with inner product

$$(f,g) = \int_\Sigma fg\, dx'.$$

With respect to this inner product, we can construct in each $\{Q_m\}$ an orthonormal basis $\{Y_{\ell m}\}$, $\ell = 1, 2, \ldots, d(m)$. In other words, for each fixed m,

$$(Y_{im}, Y_{jm}) = \int_\Sigma Y_{im} Y_{jm}\, dx' = \delta_{ij}.$$

Combining all these bases, we obtain the following result.

THEOREM. 3.8. The collection $\{Y_{\ell m}\}$, $\ell = 1, 2, \ldots, d(m)$ and $m = 0, 1, 2, \ldots$, is a complete orthonormal system on Σ.

Proof. Since, by Lemma 3.7, spherical harmonics of distinct degrees are orthogonal, $\{Y_{\ell m}\}$ is an orthonormal family. Since the continuous functions are dense in L^2, Corollary 3.6 implies that the set of all finite linear combinations of $Y_{\ell m}$ is dense in L^2. Consequently the system $\{Y_{\ell m}\}$ is complete.

Q.E.D.

Our next objective is to obtain certain bounds for these spherical harmonics $Y_{\ell m}$ and their derivatives. These bounds can be deduced from the fact that, for each m and all $x \in \Sigma$,

$$\sum_{\ell=1}^{d(m)} Y_{\ell m}^2(x) = A_m$$

where A_m is a constant depending only on m and the dimension n. In order to prove this result we introduce the concept of zonal harmonic.

LEMMA. 3.9. For each m and each point $x \in \Sigma$, there exists a spherical harmonic $Z_x \in \{Q_m\}$, called a zonal harmonic with pole x, such that for all $Q \in \{Q_m\}$

1) $$Q(x) = (Q, Z_x).$$

Moreover, for all $y \in \Sigma$,

$$2) \qquad Z_x(y) = \sum_{\ell} Y_{\ell m}(x)\, Y_{\ell m}(y)$$

where $\ell = 1, 2, \ldots, d(m)$.

Proof. Since the map $Q \to Q(x)$ is a linear functional over the (finite dimensional) Hilbert space $\{Q_m\}$, the first conclusion follows directly from Riesz Representation Theorem. Moreover, relative to the orthonormal basis $\{Y_{\ell m}\}$, $\ell = 1, \ldots, d(m)$, any Q in $\{Q_m\}$ can be written uniquely as a finite sum $Q = \sum_{\ell} a_\ell Y_{\ell m}$, where $a_\ell = a_{\ell m} = (Q, Y_{\ell m}) = \int_\Sigma Q(y)\, Y_{\ell m}(y)\, d\sigma$, $d\sigma =$ surface element on Σ.

In particular, for each fixed x,

$$(Q, Z_x) = Q(x) = \sum_{\ell} (Q(y), Y_{\ell m}(y))\, Y_{\ell m}(x) =$$

$$= (Q(y), \sum_{\ell} Y_{\ell m}(x)\, Y_{\ell m}(y)).$$

Since this holds for all Q in $\{Q_m\}$, formula 2) follows.

Q.E.D.

LEMMA. 3.10. For any rotation $u: \Sigma \to \Sigma$ and any zonal harmonic Z_x,

$$Z_{ux}(uy) = Z_x(y).$$

Proof. If the function $Q(x)$ belongs to $\{Q_m\}$ then so does the function $Q(ux)$ since the Laplacian, being associated to $P(x) = |x|^2$, is invariant under rotations.

Applying formula 1) of Lemma 3.9 to the harmonic $Q_u(x) = Q(ux)$ we have that

$$Q(ux) = Q_u(x) = \int_\Sigma Q_u(y)\, Z_x(y)\, d\sigma = \int_\Sigma Q(uy)\, Z_x(y)\, d\sigma.$$

Applying the same formula to the harmonic $Q(x)$ and then changing variable, we obtain

$$Q(ux) = \int_\Sigma Q(y)\, Z_{ux}(y)\, d\sigma = \int_\Sigma Q(uy)\, Z_{ux}(uy)\, d\sigma$$

since the measure $d\sigma$ is invariant under rotations. Hence, for all Q in $\{Q_m\}$

$$\int_\Sigma Q(uy)\, Z_x(y)\, d\sigma = \int_\Sigma Q(uy)\, Z_{ux}(uy)\, d\sigma$$

and the conclusion follows.

Q.E.D.

REMARK. If the pole $x \in \Sigma$ of a zonal harmonic Z_x lies on the axis of a rotation u, then $ux = x$, so by Lemma 3.10

$$Z_x(y) = Z_x(uy) \qquad , \text{ for all } y \in \Sigma.$$

In other words, a zonal harmonic Z_x is constant along each "parallel" relative to the pole x.

<u>LEMMA</u>. 3.11. For each m and all $x \in \Sigma$,

$$\sum_{\ell} Y_{\ell m}^2(x) = \frac{d(m)}{w_n}$$

where w_n = area of Σ.

<u>Proof</u>. By formula 2) of Lemma 3.9 we have that, for any given $x \in \Sigma$,

$$Z_x(x) = \sum_{\ell} Y_{\ell m}^2(x).$$

Similarly, for any rotation u,

$$Z_{ux}(ux) = \sum_{\ell} Y_{\ell m}^2(ux).$$

Hence, Lemma 3.10 implies that

$$\sum_{\ell} Y_{\ell m}^2(x) = \sum_{\ell} Y_{\ell m}^2(ux).$$

Since each point y of Σ is the image $y = ux$ under a suitable rotation u, we must have that

$$\sum_{\ell} Y_{\ell m}^2(x) = A_m$$

is constant for all $x \in \Sigma$.

Integrating over Σ,

$$\int_\Sigma \sum_\ell Y_{\ell m}^2(x)\, d\sigma = A_m \int_\Sigma d\sigma = A_m w_n.$$

On the other hand, since $\{Y_{\ell m}\}$, $\ell = 1, 2, \ldots, d(m)$, is an orthonormal basis,

$$\int_\Sigma \sum_\ell Y_{\ell m}^2(x)\, d\sigma = \sum_{\ell=1}^{d(m)} \int_\Sigma Y_{\ell m}(x)\, Y_{\ell m}(x)\, d\sigma = d(m).$$

Consequently, $A_m w_n = d(m)$.

Q.E.D.

We are now able to deduce the desired bounds for the $Y_{\ell m}$.

<u>THEOREM</u>. 3.12. For all $x' \in \Sigma$,

a) $$|Y_{\ell m}(x')| \le C m^{(n-2)/2}$$

where the constant C depends only on the dimension n. More generally,

b) $$\left|\left(\frac{\partial}{\partial x}\right)^\alpha [|x|^m\, Y_{\ell m}(x')]\right| \le C m^{(n-2)/2+|\alpha|}\, |x|^{m-|\alpha|}$$

where now the constant C depends only on $|\alpha|$ and n.

Proof. On account of Lemma 3.11 and Remark 3.5 d),

$$Y_{\ell m}^2(x') \le \sum_{\ell} Y_{\ell m}^2(x') = \frac{d(m)}{w_n} \le Cm^{n-2}$$

for some positive constant C depending only on n. Hence, taking square roots, we obtain estimate a).

Clearly, for $|\alpha| = 0$, estimate b) reduces to a). For the case $|\alpha| = 1$, we let $P(x) = |x|^m Y_{\ell m}(x')$ and estimate an arbitrary first partial $\frac{\partial P}{\partial x_i}$.

Since $P \in \{S_m\}$ and $\Delta \frac{\partial P}{\partial x_i} = \frac{\partial}{\partial x_i} \Delta P = 0$, $\frac{\partial P}{\partial x_i} \in \{S_{m-1}\}$ so, letting $k = m - 1$ and using Lemma 3.11, we obtain

$$\left|\frac{\partial P}{\partial x_i}(x')\right|^2 = \left|\sum_{\ell} a_\ell Y_{\ell k}(x')\right|^2 \le \{\sum_{\ell} a_\ell^2\}\{\sum_{\ell} Y_{\ell k}^2(x')\} = A\,\frac{d(k)}{w_n}$$

where $A = \{\sum_{\ell} a_\ell^2\} = \int_\Sigma \left|\frac{\partial P}{\partial x_i}\right|^2 d\sigma \le \int_\Sigma |\text{grad } P|^2\, d\sigma$.

Consequently,

$$1) \qquad \left|\frac{\partial P}{\partial x_i}(x')\right|^2 \le \frac{d(m-1)}{w_n} \int_\Sigma |\text{grad } P|^2\, d\sigma.$$

Since grad P is homogeneous of degree $m - 1$, letting $r = |x|$ and using polar coordinates we see that

$$\int_{|x|\le 1} |\text{grad } P|^2\, dx = \int_0^1 r^{2m+n-3}\, dr \int_\Sigma |\text{grad } P|^2\, d\sigma$$

and hence

$$2)\quad \int_{\Sigma} |\text{grad } P|^2 \, d\sigma = (2m + n - 2) \int_{|x|\le 1} |\text{grad } P|^2 \, dx.$$

But, since P is harmonic, $\text{div}[P \text{ grad } P] = |\text{grad } P|^2 + P\Delta P =$ $= |\text{grad } P|^2$ so, if ν = outer normal to Σ, Gauss formula yields

$$3)\quad \int_{|x|\le 1} |\text{grad } P|^2 \, dx = \int_{\Sigma} P(\text{grad } P) \cdot \nu \, d\sigma.$$

Moreover, since P is homogeneous of degree m, we have by Euler's formula that $(\text{grad } P) \cdot \nu = mP$. Consequently,

$$4)\quad \int_{\Sigma} P(\text{grad } P) \cdot \nu \, d\sigma = m \int_{\Sigma} P^2 \, d\sigma = m \int_{\Sigma} Y_{\ell m}^2 \, d\sigma = m.$$

Combining 2), 3), and 4), we conclude that

$$\int_{\Sigma} |\text{grad } P|^2 \, d\sigma = m(2m + n - 2).$$

Substituting this last formula in 1) and using Remark 3.5 d), we obtain that for all x' in Σ

$$\left|\frac{\partial P}{\partial x_i}(x')\right|^2 \le C(m - 1)^{n-2} \, m(2m + n - 2) \le C_1 m^n.$$

Therefore, since $\frac{\partial P}{\partial x_i}$ is homogeneous of degree $m - 1$, it follows that for any $i = 1, 2, \ldots, n$

$$\left|\frac{\partial P}{\partial x_i}(x)\right| \le C m^{n/2} \, |x|^{m-1}$$

which proves estimate b) for the case $|\alpha| = 1$.

Repeating this argument, we obtain estimate b) in the general case.

Q.E.D.

Returning to Theorem 3.8, it is clear that any $f \in L^2(\Sigma)$ can be represented by the Fourier series

1) $$f(x') = \sum_{\ell,m} a_{\ell m} Y_{\ell m}(x') = \sum_{m=0}^{\infty} \sum_{\ell=1}^{d(m)} a_{\ell m} Y_{\ell m}(x')$$

which, by Riesz-Fisher Theorem, converges to f in L^2 norm. The coefficients $a_{\ell m}$ are given by

2) $$a_{\ell m} = \int_{\Sigma} f(x')\, Y_{\ell m}(x')\, dx'$$

and satisfy Parseval's equality

3) $$\sum_{\ell,m} |a_{\ell m}|^2 = \int_{\Sigma} |f(x')|^2\, dx'.$$

REMARK. Any continuous function f on Σ has a continuous extension $\tilde{f}(x) = f(x/|x|)$ to $E^n - \{0\}$ which is homogeneous of degree zero. Conversely, any $g \in C(E^n_o)$ which is homogeneous of degree zero is of the form $g = \tilde{f}$ where f is its restriction to the unit sphere Σ. In general, we shall not distinguish between f and $\tilde{f}$, and when we write $f \in C^\infty(\Sigma)$ we mean that $\tilde{f} \in C^\infty(E^n_o)$.

If $f \in C^{\infty}(\Sigma)$ then, as is well known, the Fourier series in 1) converges absolutely and uniformly to f. In fact, we are going to prove more, namely that a necessary and sufficient condition for a sequence $\{a_{\ell m}\}_{\ell=1,2,\ldots,d(m)}^{m=0,1,2,\ldots}$ to be the "harmonic Fourier coefficients" of a function $f \in C^{\infty}(\Sigma)$ is that the sequence be rapidly decreasing. This last result is a consequence of the analogue of formula 2) above, which we shall prove in Theorem 3.15 below. First we introduce an operator L, which preserves the degree of homogeneity of a function, given by

$$Lf = |x|^2 \, \Delta f.$$

Clearly, $Lf = \Delta f$ on Σ.

<u>LEMMA</u>. 3.13. For all $Y_{\ell m}$ and all integers $r \geq 0$

$$L^r Y_{\ell m} = (-m)^r \, (m + n - 2)^r \, Y_{\ell m}.$$

<u>Proof</u>. Letting $P(x) = |x|^m \, Y_{\ell m}(x')$ be the corresponding solid harmonic, we have that

1) $$LY_{\ell m} = |x|^2 \, \Delta Y_{\ell m} = |x|^2 \, \Delta[|x|^{-m} \, P(x)].$$

From the formulas

$$\Delta(fg) = f\Delta g + 2(\text{grad } f) \cdot (\text{grad } g) + g\Delta f$$

$$\text{grad } |x|^k = k|x|^{k-2}\, x$$

$$\Delta|x|^k = k(k + n - 2)\, |x|^{k-2}$$

$$x \cdot (\text{grad } P) = mP$$

and the harmonicity of P, it follows that

$$\Delta[|x|^{-m} P(x)] = 2(\text{grad } |x|^{-m})\cdot(\text{grad } P) + (-m)(-m + n - 2)|x|^{-m-2} P(x) =$$

$$= 2(-m)|x|^{-m-2}\, x\cdot(\text{grad } P) + (-m)(-m + n - 2)|x|^{-m-2} P(x) =$$

$$= 2(-m^2)|x|^{-m-2} P(x) + (-m)(-m + n - 2)|x|^{-m-2} P(x) =$$

$$= (-m)(m + n - 2)|x|^{-m-2} P(x).$$

Hence, substituting in 1), we see that

$$LY_{\ell m} = (-m)(m + n - 2)\, Y_{\ell m}$$

and, by iteration, we obtain the conclusion.

Q.E.D.

<u>LEMMA</u>. 3.14. If $f, g \in C^{2r}(E^n_o)$ and are homogeneous of degree zero, then

$$\int_\Sigma fL^r g\, d\sigma = \int_\Sigma gL^r f\, d\sigma.$$

Proof. Using polar coordinates we see that for some constant $C \neq 0$

$$\int_{1\le|x|\le 2} (f\Delta g - g\Delta f)\, dx = C \int_\Sigma (f\Delta g - g\Delta f)\, d\sigma.$$

On the other hand, by Green's formula,

$$\int_{1\le|x|\le 2} (f\Delta g - g\Delta f)\, dx = \left\{\int_{|x|=2} - \int_{|x|=1}\right\}\left(f\frac{\partial g}{\partial \nu} - g\frac{\partial f}{\partial \nu}\right) = 0$$

because $\frac{\partial}{\partial\nu} = \frac{\partial}{\partial|x|}$ is the radial derivative but f and g are constant along radii. Hence

$$\int_\Sigma (f\Delta g - g\Delta f)\, d\sigma = 0.$$

Since, on Σ, $Lf = \Delta f$ and $Lg = \Delta g$, we see that

$$\int_\Sigma fLg\, d\sigma = \int_\Sigma gLf\, d\sigma.$$

Replacing g by Lg in this expression, it follows that

$$\int_\Sigma fL^2 g = \int_\Sigma (Lg)(Lf) = \int_\Sigma gL^2 f$$

and, repeating this process, we obtain the conclusion.

Q.E.D.

THEOREM. 3.15. Let $f \in C^\infty(\Sigma)$ and $\{a_{\ell m}\}$ be its Fourier coefficients with respect to $\{Y_{\ell m}\}$. Then, for every integer $r \geq 0$,

$$a_{\ell m} = (-m)^{-r} (m + n - 2)^{-r} \int_\Sigma Y_{\ell m} L^r f \, d\sigma.$$

Proof. By the preceding Lemmas

$$a_{\ell m} = \int_\Sigma f Y_{\ell m} \, d\sigma = (-m)^{-r} (m + n - 2)^{-r} \int_\Sigma f L^r Y_{\ell m} \, d\sigma =$$

$$= (-m)^{-r} (m + n - 2)^{-r} \int_\Sigma Y_{\ell m} L^r f \, d\sigma.$$

Q.E.D.

THEOREM. 3.16. Let $f \in C^\infty(\Sigma)$ and $\{a_{\ell m}\}$ be its Fourier coefficients with respect to $\{Y_{\ell m}\}$. Then, for every integer $r \geq 0$,

(*) $$\sum_{\ell, m} m^r |a_{\ell m}| < \infty.$$

Conversely, given any family of constants $\{a_{\ell m}\}_{\ell=1,2,\ldots,d(m)}^{m=0,1,2,\ldots}$ which satisfies (*), there exists an $f \in C^\infty(\Sigma)$ such that

$$f(x') = \sum_{\ell, m} a_{\ell m} Y_{\ell m}(x').$$

Proof. Clearly, it suffices to prove (*) for all r sufficiently large. By Theorem 3.15,

$$|a_{\ell m}| = m^{-r}(m + n - 2)^{-r} \left| \int_\Sigma Y_{\ell m} L^r f \, d\sigma \right|$$

so, using Schwarz's inequality and the normality of $Y_{\ell m}$, we obtain

$$|a_{\ell m}| \leq m^{-2r} \left\{ \int_\Sigma |L^r f|^2 \, d\sigma \right\}^{1/2} = C_r m^{-2r}.$$

Hence, on account of the estimate $d(m) \leq Cm^{n-2}$, it follows that

$$\sum_{\ell,m} m^r |a_{\ell m}| \leq C_r \sum_{m=1}^{\infty} d(m)\, m^{-r} \leq CC_r \sum_{m=1}^{\infty} m^{n-2-r} < \infty$$

for all $r > n - 1$.

Conversely, suppose that $\{a_{\ell m}\}$ satisfies (*) for any $r \geq 0$. Since by Theorem 3.12 we have the estimate $|Y_{\ell m}(x')| \leq Cm^{(n-2)/2}$ with C independent of m and x', it follows from (*) that the series $\sum_{\ell,m} a_{\ell m} Y_{\ell m}(x')$ converges absolutely and uniformly to a continuous function f on Σ.

Repeating this argument using estimates b) of Theorem 3.12, we see that the series $\sum_{\ell,m} a_{\ell m} \left(\frac{\partial}{\partial x}\right)^{\alpha} Y_{\ell m}(x')$ is absolutely and uniformly convergent to the continuous function $\left(\frac{\partial}{\partial x}\right)^{\alpha} f$ on Σ. Hence $f \in C^{\infty}(\Sigma)$.

Q.E.D.

We conclude this chapter with some considerations on Fourier transforms, motivated by the fact that in the next chapter we need to know the Fourier transform of $\text{p.v.} Y_{\ell m}(x')\, |x|^{-n}$, $m \geq 1$.

We begin with the following lemma which generalizes the familiar fact that the function $\exp(-\pi\, |x|^2)$ is equal to its

Fourier transform.

LEMMA. 3.17. If $P \in \{S_m\}$, then

$$F[P(x) \exp(-\pi|x|^2)] = (-i)^m P(x) \exp(-\pi|x|^2).$$

Proof. (The following inductive proof is due to C. Lemoine.) If $m = 0$, hence P is a constant, the equality is well known. Since if $P \in \{S_m\}$, $m \geq 1$, then $\frac{\partial P}{\partial x_k} \in \{S_{m-1}\}$ for any $k = 1, 2, \ldots, n$, we may assume that the formula holds for each $p^{(k)} = \frac{\partial P}{\partial x_k}$.

Since P is homogeneous of degree m, $mP = x \cdot \operatorname{grad} P$ so, using the inductive assumption, we obtain

$$\begin{aligned} F[P(x) \exp(-\pi\,|x|^2)] &= \frac{1}{m} \sum_k F[x_k P^{(k)}(x) \exp(-\pi|x|^2)] = \\ &= \frac{1}{m} \cdot \frac{i}{2\pi} \sum_k \frac{\partial}{\partial x_k} F[P^{(k)}(x) \exp(-\pi|x|^2)] = \\ &= \frac{1}{m} \cdot \frac{(-i)^m}{-2\pi} \sum_k \frac{\partial}{\partial x_k} [P^{(k)}(x) \exp(-\pi|x|^2)] = \\ &= \frac{1}{m} \cdot \frac{(-i)^m}{-2\pi} \sum_k P^{(k)}(x)(-2\pi x_k) \exp(-\pi|x|^2) = \\ &= (-i)^m P(x) \exp(-\pi|x|^2) \end{aligned}$$

where, in the next to last line, we have used the fact that $\sum_k \frac{\partial}{\partial x_k} P^{(k)}(x) = \Delta P(x) = 0$ since P is harmonic.

Q.E.D.

Consider a spherical harmonic $Q \in \{Q_m\}$, $m \geq 1$. Clearly, $Q \in L^2(\Sigma)$ and has mean value zero there, because

$$\int_\Sigma Q(x')\,dx' = \int_\Sigma 1 \cdot Q(x')\,dx' = 0$$

since constants are harmonics of degree zero. Hence, the function $Q(x')|x|^{-n}$ is a singular convolution kernel and, by Theorem 2.1.6, p.v.$Q(x')|x|^{-n}$ is a temperate distribution whose Fourier transforms is a bounded function homogeneous of degree zero and with mean value zero on Σ. This function, as we see in the next theorem which generalizes the formula $F[\text{p.v.}\frac{x_k}{|x|^{n+1}}] = \gamma \frac{x_k}{|x|}$ of the Riesz transforms, is in fact a constant multiple of $Q(x')$.

THEOREM. 3.18. Let $Q \in \{Q_m\}$, $m \geq 1$. Then

$$F[\text{p.v.}Q(x')|x|^{-n}] = \gamma_m\, Q(x')$$

where

$$\gamma_m = (-i)^m\, \pi^{n/2}\, \Gamma\left(\frac{m}{2}\right)\Big/\Gamma\left(\frac{n+m}{2}\right).$$

Proof. Let $V = F[\text{p.v.}Q(x')|x|^{-n}]$. As we just noted, V is a bounded function homogeneous of degree zero and with mean value zero on Σ. For any solid harmonic $P \in \{S_k\}$, $k > 0$, we see by Plancherel's formula and Lemma 3.17 that

1)

$$\langle \text{p.v.}Q(x')|x|^{-n},\ \ P(x)\exp(-\pi|x|^2)\rangle = \langle V(x),(-i)^k\, P(x)\exp(-\pi|x|^2)\rangle .$$

In polar coordinates, the left side of 1) equals

$$\int_0^\infty r^{k-1}\, e^{-\pi r^2}\, dr \int_\Sigma Q(x')\, P(x')\, dx' = C_1 \int_\Sigma Q(x')\, P(x')\, dx'$$

and the right side of 1) equals

$$i^k \int_0^\infty r^{k+n-1}\, e^{-\pi r^2}\, dr \int_\Sigma V(x')\, P(x')\, dx' = C_2 \int_\Sigma V(x')\, P(x')\, dx' .$$

Hence, letting $C_1/C_2 = C \neq 0$, we obtain

2) $$\int_\Sigma V(x')\, P(x')\, dx' = C \int_\Sigma Q(x')\, P(x')\, dx'.$$

Moreover, 2) holds even when $P \in \{S_o\}$ since both Q and V have mean value zero on Σ. Accordingly, we see that Q and V have the same orthogonal complement in $L^2(\Sigma)$ and hence for some constant γ_m we must have

$$V(x') = \gamma_m\, Q(x').$$

Taking $P(x) = Q(x')|x|^m$ in the preceding argument, then $k = m$ and $\gamma_m = C = C_1/C_2$, so

$$\gamma_m\, i^m \int_0^\infty r^{m+n-1}\, e^{-\pi r^2}\, dr = \int_0^\infty r^{m-1}\, e^{-\pi r^2}\, dr$$

and we can calculate γ_m as in the proof of (9) in Proposition 2.1.1.

Q.E.D.

REMARK. It is easy to see from the definition of γ_m that

$$\left|\gamma_m^{-1}\right| \sim Cm^{n/2}$$

as $m \to \infty$, where the constant C depends only on the dimension n.

CHAPTER IV

SINGULAR INTEGRAL OPERATORS

We shall study here a class of singular integral operators which contains, as a most important special case, the singular convolution operators defined in Chapter II. Basically, the new feature is that the singular convolution kernels $k(z)$ of Definition 2.1.3. are now replaced by kernels $k(x,z) = k(x,z')|z|^{-n}$ which are assumed to depend somewhat regularly on the parameter x and, as functions of z, to be of class C^∞ on $|z| > 0$. These regularity assumptions with respect to both x and z can be relaxed somewhat, but the price for greater generality (particularly with regard to x) is a considerable loss of simplicity.

In the spaces L^p_k considered here, k will always be an integer. Let $\beta \geq 0$ be a real number and let $b = [\beta]$ be its integral part.

DEFINITION. 4.0. We denote by B_β the class of bounded function whose distribution derivatives of order $\leq b = [\beta]$ coincide with bounded functions and such that the derivatives of order b satisfy a uniform Hölder condition of order $\beta - b$.

In other words, if $f \in B_\beta$ and $|\alpha| = b$ then there exists a positive constant M_α such that, for all x and y in E^n,

$$|D^{\alpha}f(y) - D^{\alpha}f(x)| \leq M_{\alpha}|x - y|^{\beta-b}.$$

We define the norm of any $f \in B_{\beta}$ by the formula

$$\|f\|_{\beta} = \max \left\{ \sup_{x \in E^n} |D^{\alpha}f(x)|, \quad 0 \leq |\alpha| \leq b; \quad M_{\alpha}, \quad |\alpha| = b \right\}$$

where M_{α} are the smallest constants for which the Hölder estimates hold.

REMARK. When $\beta = b$ is an integer, $B_{\beta} = B_b$ coincides with the Sobolev space L_b^{∞}. However, the Hölder continuity assumption turns out to be very convenient in the proof of Theorem 4.9 below.

<u>THEOREM</u>. 4.1. (a) If $f, g \subset B_{\beta}$ then $f + g$ and fg are also in B_{β}. Moreover, $\|f + g\|_{\beta} \leq \|f\|_{\beta} + \|g\|_{\beta}$ and $\|fg\|_{\beta} \leq C\|f\|_{\beta}\|g\|_{\beta}$ where C depends only on β.

(b) If $|k| \leq \beta$ then L_k^p is a B_{β}-module, that is if $f \in B_{\beta}$ and $g \in L_k^p$ then $(fg) \in L_k^p$ and $\|fg\|_{p,k} \leq C\|f\|_{\beta}\|g\|_{p,k}$ where C depends only on β.

<u>Proof</u>. Part (a) is straight forward. Part (b) is obvious for $k = 0$. Let $0 < k \leq \beta$. If $g \in L_k^p$ then

$$\|fg\|_{p,k}^2 = \sum_{|\alpha| \leq k} \|D^{\alpha}(fg)\|_p^2 \leq \sum_{|\alpha| \leq k} \sum_{|\mu| + |\nu| = |\alpha|} C\|(D^{\mu}f)(D^{\nu}g)\|_p^2 \leq$$

$$\leq C\|f\|_{\beta}^2 \|g\|_{p,k}^2 .$$

If $g \in L^p_{-k}$, $1 \leq p < \infty$ and $p' = p/(p-1)$ is the conjugate exponent, then, for all $h \in L^{p'}_k$ we obtain, by Theorem 2.2.12 and what we have already proved, that

$$|(h,fg)| = |(fh,g)| \leq \|fh\|_{p',k} \|g\|_{p,-k} \leq$$

$$\leq C\|f\|_\beta \|h\|_{p',k} \|g\|_{p,-k} .$$

Consequently, $(fg) \in L^p_{-k}$ and $\|fg\|_{p,-k} \leq C\|f\|_\beta \|g\|_{p,-k}$.

Q.E.D.

DEFINITION. 4.2. An operator $f \to Kf$ of the form

$$(*) \qquad [Kf](x) = a(x)\, f(x) + \lim_{\varepsilon\to 0} \int_{|x-y|>\varepsilon} k(x,x-y)\, f(y)\, dy$$

is said to be a singular integral operator, if, for each fixed $x \in E^n$, $k(x,z)$ is homogeneous of degree $-n$ in z, of class C^∞ on $|z| > 0$, and has mean value zero on $|z| = 1$.

Given these properties of $k(x,z)$, for each fixed x, the principal value Fourier transform with respect to z, denoted by

$$h(x,\zeta) = k(x,\hat{\zeta}) = \lim_{\varepsilon\to 0} \int_{\varepsilon<|z|<1/\varepsilon} e^{-2\pi i(\zeta,z)}\, k(x,z)\, dz,$$

exists and defines a function homogeneous of degree zero in ζ, of class C^∞ on $|\zeta| > 0$, with mean value zero on $|\zeta| = 1$. (Cf. Theorem 2.1.7, and also [6], Chapter IV, Theorem 4).

<u>DEFINITION</u>. 4.3. Given a singular integral operator K, the function $\sigma(K)$ given by

$$(**) \quad \sigma(K)(x,\zeta) = a(x) + h(x,\zeta) = a(x) + k(x,\hat{\zeta})$$

is called the <u>symbol</u> of K.

Since $h(x,\zeta)$ has mean value zero on $|\zeta| = 1$, it is clear from (**) that $a(x)$ is precisely the mean value of the symbol on $|\zeta| = 1$. Moreover, any function $s(x,\zeta)$, which is homogeneous of degree zero in ζ and of class C^∞ on $|\zeta| > 0$, must be the symbol, $s = \sigma(K)$, of some singular integral operator K. If fact, fixing x and letting $m(x)$ be the mean vlaue of $s(x,\zeta)$ on $|\zeta| = 1$, we can write $s(x,\zeta) = m(x) + [s(x,\zeta) - m(x)]$ and apply the second part of Theorem 2.1.7 to the function $s(x,\zeta) - m(x)$, as a function of ζ.

<u>DEFINITION</u>. 4.4. We shall say that K is a <u>B_β singular integral operator</u> if, for each ζ on $|\zeta| = 1$ and any α with $0 \le |\alpha| \le 2n$, the functions of x

$$\left(\frac{\partial}{\partial\zeta}\right)^\alpha \sigma(K)$$

belong to B_β. We define the norm $\|K\|_\beta$, of a B_β singular integral operator K, to be

$$\|K\|_\beta = \max_{0\le|\alpha|\le 2n} \left\{ \sup_{|\zeta|=1} \left\| \left(\frac{\partial}{\partial\zeta}\right)^\alpha \sigma(K)(x,\zeta) \right\|_\beta \right\}.$$

We note, in particular, that for any B_β singular integral operator K the functions of x

$$a(x), \quad k(x,z) \quad \text{and} \quad h(x,\zeta),$$

of formulas (*) and (**), all belong to the class B_β.

THEOREM. 4.5. Let K be a B_β singular integral operator. If $f \in L^p_k$, $1 < p < \infty$ and $0 \leq k \leq \beta$, then

$$\lim_{\varepsilon \to 0} \int_{|x-y|>\varepsilon} k(x,x-y)\, f(y)\, dy$$

exists as a limit in L^p_k norm. Moreover, for any integer k such that $|k| \leq \beta$, we have

$$\|Kf\|_{p,k} \leq C\|K\|_\beta \, \|f\|_{p,k}$$

where the constant C depends only on p and β.

Proof. Let $\{Y_{\ell m}\}^{m=0,1,2,\ldots}_{\ell=1,2,\ldots,d(m)}$ be a complete orthonormal system of spherical harmonics of degree m. By Remark 3.5 d) we have the estimate

$$d(m) \leq Cm^{(n-2)} \tag{1}$$

for some constant $C > 0$ independent of m. Moreover, from (a) of Theorem 3.12, we have that for all $z' \in \Sigma$

$$(2) \qquad |Y_{\ell m}(z')| \leq Cm^{(n-2)/2}$$

for some new constant C independent of ℓ and m.

If $F(z)$ is of class C^∞ on $|z| > 0$ and homogeneous of degree zero, it can be expanded in the convergent Fourier series

$$F(z') = c_o + \sum_{m=1}^{\infty} \sum_{\ell=1}^{d(m)} c_{\ell m} Y_{\ell m}(z')$$

and from Theorem 3.15, letting $(LF)(z) = |z|^2(\Delta F)(z)$, we can represent the coefficients $c_{\ell m}$ by the formula

$$(3) \qquad c_{\ell m} = (-1)^r m^{-r}(m + n - 2)^{-r} \int_\Sigma (L^r F)(z') Y_{\ell m}(z') dz'$$

which holds for any integer $r \geq 0$.

Furthermore, from Theorem 3.18, we recall that for all $m \geq 1$

$$(4) \qquad F[\text{p.v.}|z|^{-n} Y_{\ell m}(z')](\zeta) = \gamma_m Y_{\ell m}(\zeta')$$

for some constant γ_m dependent also on n, for which we have the estimate

$$(5) \qquad \left|\gamma_m^{-1}\right| \leq Cm^{n/2}$$

where the constant C is independent of m.

Consider now the function $|z|^n \, k(x,z)$ as a function of z only, keeping x fixed for the moment. This function of z is homogeneous of degree zero, of class C^∞ on $|z| > 0$, and has mean value zero on $|z| = 1$. Hence, letting $\sum_{\ell,m} = \sum_{m=1}^{\infty} \sum_{\ell=1}^{d(m)}$, we have the expansion

$$|z|^n \, k(x,z) = \sum_{\ell,m} a_{\ell m} Y_{\ell m}(z')$$

that is,

$$(6) \qquad k(x,z) = \left\{ \sum_{\ell,m} a_{\ell m} Y_{\ell m}(z') \right\} |z|^{-n} .$$

Similarly, we can expand $h(x,\zeta) = \sigma(K) - a(x)$, which is homogeneous of degree zero in ζ, in the form

$$(7) \qquad h(x,\zeta) = \sum_{\ell,m} b_{\ell m} Y_{\ell m}(\zeta').$$

Now, if we let x vary, we note that the coefficients $a_{\ell m}$ and $b_{\ell m}$ are functions of x in the class B_β.

According to formula (3), for any integer $r \geq 0$, we have

$$b_{\ell m}(x) = (-1)^r \, m^{-r}(m + n - 2)^{-r} \int_\Sigma (L^r h)(x,\zeta') \, Y_{\ell m}(\zeta') \, d\zeta'$$

where $L = |\zeta|^2 \, \Delta_\zeta$ and Δ_ζ is the Laplacian in ζ. Since the variable x appears only as a parameter in the preceding integral, we can differentiate with respect to x under the integral sign. Moreover, bounding the increments in x of the derivatives $\left(\frac{\partial}{\partial x}\right)^{\alpha}$ of order $b = [\beta]$, we obtain, by definition of $\|K\|_\beta$, that for any $r \leq n$

$$\|b_{\ell m}\|_\beta \leq Cm^{-2r} \|K\|_\beta \int_\Sigma |Y_{\ell m}(\zeta')| \, d\zeta'$$

where the estimate is sharpest when $r = n$. So, letting $r = n$ and applying Schwarz's inequality to the last integral, we have

$$(8) \qquad \|b_{\ell m}\|_\beta \leq Cm^{-2n} \|K\|_\beta$$

since the $Y_{\ell m}$ are normal.

Furthermore, it is immediate that we also have the bound

$$(8') \qquad \|a(x)\|_\beta \leq C\|K\|_\beta$$

and that the constants C in (8) and (8') depend only on β and the dimension n.

Combining the obvious inequality $|b_{\ell m}(x)| \leq \|b_{\ell m}\|_\beta$ with estimates (1), (2) and (8), we see that

$$\sum_{\ell=1}^{d(m)} |b_{\ell m}(x)||Y_{\ell m}(\zeta')| \leq C\|K\|_\beta \, m^{-2n} \, m^{(n-2)/2} \, m^{(n-2)} =$$

$$= C\|K\|_\beta \, m^{-3-n/2}$$

for some constant $C > 0$ depending only on β and n. Hence, the series expansion (7) of $h(x,\zeta) = \sigma(K)(x,\zeta) - a(x)$ is absolutely and uniformly convergent.

Now, for any u in $\mathcal{D}$, by Plancherel's theorem and the definition of $h(x,\zeta)$, we have that

$$\lim_{\varepsilon\to 0}\int_{|z|>\varepsilon} k(x,z)\,\overline{u(z)}\,dz = \int h(x,\zeta)\,\overline{\hat{u}(\zeta)}\,d\zeta =$$

$$= \int \Big\{\sum_{\ell,m} b_{\ell m}(x)\,Y_{\ell m}(\zeta')\Big\}\,\overline{\hat{u}(\zeta)}\,d\zeta =$$

$$= \sum_{\ell,m} b_{\ell m}(x)\,\gamma_m^{-1}\int \gamma_m Y_{\ell m}(\zeta')\,\overline{\hat{u}(\zeta)}\,d\zeta$$

where termwise integration is justified by the uniform convergence of the series expansion of h. Consequently, by formula (4) and Plancherel's theorem, we deduce that

$$\lim_{\varepsilon\to 0}\int_{|z|>\varepsilon} k(x,z)\,\overline{u(z)}\,dz =$$

$$= \sum_{\ell,m} b_{\ell m}(x)\,\gamma_m^{-1}\lim_{\varepsilon\to 0}\int_{|z|>\varepsilon} |z|^{-n}\,Y_{\ell m}(z')\,\overline{u(z)}\,dz =$$

$$= \sum_{\ell,m} b_{\ell m}(x)\,\gamma_m^{-1}\left\langle \text{p.v.}|z|^{-n}\,Y_{\ell m}(z'),\ \overline{u(z)}\right\rangle.$$

From the proof of Lemma 2.1.5 and Schwarz's inequality, we have that for some constants C, depending on u but not on ℓ or m,

$$|\langle \text{p.v.} |z|^{-n} Y_{\ell m}(z'), \bar{u} \rangle| \le C \int_{\Sigma} |Y_{\ell m}(z')| \, dz' \le C$$

so, taking into account formulas (1), (5) and (8), we conclude that the last series converges absolutely and uniformly. Hence,

$$\lim_{\varepsilon \to 0} \int_{|z|>\varepsilon} k(x,z) \overline{u(z)} \, dz =$$

$$= \lim_{\varepsilon \to 0} \int_{|z|>\varepsilon} \left\{ \sum_{\ell,m} b_{\ell m}(x) \gamma_m^{-1} Y_{\ell m}(z') \right\} |z|^{-n} \overline{u(z)} \, dz$$

and, since this expression holds for all $u \in \mathcal{D}$, it follows that

$$(9) \qquad k(x,z) = \left\{ \sum_{\ell,m} b_{\ell m}(x) \gamma_m^{-1} Y_{\ell m}(z') \right\} |z|^{-n}$$

Comparing (9) with (6) and using estimates (5) and (8), we see that

$$(9') \qquad \begin{cases} a_{\ell m}(x) = b_{\ell m}(x) \gamma_m^{-1} \\ \\ \|a_{\ell m}\|_{\beta} \le C m^{-3n/2} \|K\|_{\beta}. \end{cases}$$

Furthermore, since $|a_{\ell m}(x)| \le \|a_{\ell m}\|_{\beta}$, estimates (1), (2) and (9') imply that

$$\sum_{\ell=1}^{d(m)} |a_{\ell m}(x)| |Y_{\ell m}(z')| \le C \|K\|_{\beta} \, m^{-3}.$$

Consequently, the series expansion (9) of $k(x,z)$ is absolutely and uniformly convergent.

Let us now define the "truncated" operators K_ε and $G_\varepsilon^{\ell m}$ by setting

$$[K_\varepsilon f](x) = a(x)\, f(x) + \int_{|x-y|>\varepsilon} k(x,x-y)\, f(y)\, dy$$

$$[G_\varepsilon^{\ell m} f](x) = \int_{|x-y|>\varepsilon} |x-y|^{-n}\, Y_{\ell m}(x-y)\, f(y)\, dy$$

By Hölder's inequality, the integrals exist for all $f \in L^p$, $1 < p < \infty$. Moreover, expanding $k(x,x-y)$ as in (9) and integrating term by term, we have the representation

$$(10) \qquad K_\varepsilon f = a(x)\, f + \sum_{\ell,m} a_{\ell m}(x)\, G_\varepsilon^{\ell m} f.$$

Applying Theorem 2.1.4, with $q = 2$, we obtain that $\|G_\varepsilon^{\ell m} f\|_p \leq$ $\leq C\|f\|_p$, where $C = A_{p,2}\left\{\int_\Sigma |Y_{\ell m}|^2\, d\sigma\right\}^{1/2} = A_{p,2}$ depends only on p and the dimension n, and that, as $\varepsilon \to 0$, $G_\varepsilon^{\ell m} f$ converges in L^p norm to a limit denoted by $G^{\ell m} f$. More generally, by Theorems 2.2.15 and 2.2.17, we have that, for any integer k, if $f \in L_k^p$ then

$$(11) \qquad \|G_\varepsilon^{\ell m} f\|_{p,k} \leq C\|f\|_{p,k}$$

where C depends on p and n only, and $G_\varepsilon^{\ell m} f \to G^{\ell m} f$ in L_k^p norm, as $\varepsilon \to 0$, for any integer $k \geq 0$.

Now, let $f \in L^p_k$, $1 < p < \infty$ and $0 \le |k| \le \beta$. Then, by Theorem 4.1 and formulas (10) and (11), it follows that for various constants $C > 0$ (depending at most on β, p and n)

$$\|K_\varepsilon f\|_{p,k} \le C\|a\|_\beta\|f\|_{p,k} + C\|f\|_{p,k} \sum_{\ell,m} \|a_{\ell m}\|_\beta.$$

But, by (1) and (9'),

$$\sum_{\ell=1}^{d(m)} \|a_{\ell m}\|_\beta \le C\|K\|_\beta\, m^{-3n/2}\, m^{(n-2)}$$

$$\sum_{\ell,m} \|a_{\ell m}\|_\beta = \sum_{m=1}^{\infty} \sum_{\ell=1}^{d(m)} \|a_{\ell m}\|_\beta \le C\|K\|_\beta \sum_{m=1}^{\infty} m^{\;2-n/2}$$

and, by (8'), $\|a\|_\beta \le C\|K\|_\beta$. Therefore, for any $\varepsilon > 0$, we conclude that

$$\|K_\varepsilon f\|_{p,k} \le C\|K\|_\beta\|f\|_{p,k} \tag{12}$$

for all $f \in L^p_k$, $1 < p < \infty$ and $0 \le |k| \le \beta$, where the constant C depends only on p and β (and the dimension n).

Moreover, by (10) and the definition of K, it follows that, as $\varepsilon \to 0$, $K_\varepsilon f \to Kf$ in L^p_k norm, and by continuity of the norm, Kf also satisfies estimate (12).

Q.E.D.

We note that, in the course of the preceding proof, we have

also established the following result.

PROPOSITION. 4.6. Let K be a B_β singular integral operator with kernel $k(x,z) = \left\{\sum_{\ell,m} a_{\ell m}(x)\, Y_{\ell m}(z')\right\} |z|^{-n}$ where $\{Y_{\ell m}\}$ is a complete orthonormal system of spherical harmonics, and let $G^{\ell m}$ be the Giraud operators

$$[G^{\ell m}f](x) = \lim_{\varepsilon\to 0} \int_{|x-y|>\varepsilon} |x - y|^{-n}\, Y_{\ell m}(x - y)\, f(y)\, dy.$$

Then, the series of operators $a(x) + \sum_{\ell,m} a_{\ell m}(x)\, G^{\ell m}$ converges to the operator K, on the spaces L^p_k where $1 < p < \infty$ and $0 \le |k| \le \beta$.

Given a B_β singular integral operator K, where

$$[Kf](x) = \lim_{\varepsilon\to 0} [K_\varepsilon f](x) = a(x)\, f(x) + \lim_{\varepsilon\to 0} \int_{|x-y|>\varepsilon} k(x, x - y)\, f(y)\, dy,$$

we can define its adjoint K^* as the singular integral operator

$$(*) \qquad [K^*f](x) = \lim_{\varepsilon\to 0} [K^*_\varepsilon f](x) =$$

$$= \bar{a}(x)\, f(x) + \lim_{\varepsilon\to 0} \int_{|x-y|>\varepsilon} \bar{k}(y, y - x)\, f(y)\, dy$$

where $\bar{a}$ and $\bar{k}$ are the complex conjugates of the functions a and k, respectively. Equivalently, we could define K^* on $\mathcal{D}$ by the usual formula

$$(\dagger) \qquad \int [Kf]\,\bar{g}\,dx = \int f\overline{[K^*g]}\,dx$$

for all $f, g \in \mathcal{D}$, and verify, interchanging the order of integration, that K^* has indeed the form (*).

Since, by Theorem 4.5, K is continuous on L^p_k for all $1 < p < \infty$ and $|k| \le \beta$, the same result holds, by duality, for the adjoint operator K^*. Thus, if K^* is defined on $\mathcal{D}$ by (†), then it can be defined by continuity on the whole space L^p_k. Hence, rewriting (†) in the form

$$\langle Kf, \bar{g} \rangle = \langle f, \overline{K^*g} \rangle ,$$

this formula is valid for all $f \in L^p_k$ and $g \in L^{p'}_{-k}$, where $1 < p < \infty$, $p' = p/(p-1)$, and $|k| \le \beta$.

Unfortunately, as we see from (*), K^* fails to be a B_β singular integral operator unless, for example, K is actually a singular convolution operator, that is the symbol $\sigma(K)(x,z) = \sigma(K)(z)$ is independent of x. Similarly, the usual composition K_2K_1 of two B_β singular integral operators is not an operator of the same type, in general.

However, this situation can be partially remedied by defining a "pseudo-product" and a "pseudo-adjoint" of B_β singular integral operators, which will again be operators of the same type and which "will not differ too much" from the usual composition and adjoint.

DEFINITION. 4.7. If K, K_1 and K_2 are B_β singular integral operators we define the pseudo-adjoint $K^{\#}$ and the pseudo-product $K_1 \circ K_2$ to be the B_β singular integral operators with symbols

(i) $$\sigma(K^{\#}) = \overline{\sigma(K)}$$

(ii) $$\sigma(K_1 \circ K_2) = \sigma(K_1)\, \sigma(K_2).$$

REMARKS. From definition 4.4 and the comment preceding it, it is clear that the functions $\overline{\sigma(K)}$ and $\sigma(K_1)\, \sigma(K_2)$ are in fact symbols of B_β singular integral operators and the uniqueness of these operators is also clear. Hence, $K^{\#}$ and $K_1 \circ K_2$ are well-defined. Moreover, we note that $K^{\#} = K$ if and only if $\sigma(K)$ is real-valued, and that $K_1 \circ K_2 = K_2 \circ K_1$, so (unlike the usual composition of operators) the pseudo-product is commutative.

PROPOSITION. 4.8. Let K, K_1 and K_2 be B_β singular integral operators with symbols independent of x. Then,

(a) $K^* = K^{\#}$ and $K_1 \circ K_2 = K_1 K_2 = K_2 K_1$.

(b) If $\sigma(K)$ never vanishes, then K has an inverse which is again a B_β singular integral operator.

Proof. Let $f \in \mathcal{D}$ and $\hat{f}$ be its Fourier transform. According to formula (10) in the proof of Theorem 4.5,

$$K_\varepsilon f = af + \sum_{\ell,m} a_{\ell m} G_\varepsilon^{\ell m} f$$

where now a and $a_{\ell m}$ are constants, since they appear in the harmonic expansion of $\sigma(K)$. By definition of $G_\varepsilon^{\ell m}$ and formula (4), taking Fourier transforms in the preceding expression and letting $\varepsilon \to 0$, we obtain

$$(Kf)^\wedge(\zeta) = a\hat{f}(\zeta) + \sum_{\ell,m} a_{\ell m} \gamma_m Y_{\ell m}(\zeta') \hat{f}(\zeta).$$

Therefore, on account of (7) and (9'),

$$\text{(i)} \qquad (Kf)^\wedge = \sigma(K) \hat{f} .$$

Moreover, by continuity of the operations on both sides, (i) holds for $f \in L^2$.

From this formula it follows that for all $g \in \mathcal{D}$

$$\langle (Kf)^\wedge, \bar{\hat{g}} \rangle = \langle \sigma(K)\hat{f}, \bar{\hat{g}} \rangle = \langle \hat{f}, \overline{\overline{\sigma(K)}\, \hat{g}} \rangle.$$

On the other hand, by Plancherel's formula and the definition of K^*,

$$\langle (Kf)^\wedge, \bar{\hat{g}} \rangle = \langle Kf, \bar{g} \rangle = \langle f, \overline{K^* g} \rangle =$$

$$= \langle \hat{f}, \overline{(K^* g)^\wedge} \rangle.$$

Consequently, $(K^*g)\hat{} = \overline{\sigma(K)}\,\hat{g}$. Since by (i) and the definition of $K^\#$ we also have that $(K^\# g)\hat{} = \overline{\sigma(K)}\,\hat{g}$, taking inverse Fourier transforms we see that $K^*g = K^\# g$, for all $g \in \mathcal{D}$.

Now, from (i) and the definition of $K_1 \circ K_2$, it follows that, for all $f \in \mathcal{D}$,

$$[(K_1 \circ K_2)f]\hat{} = \sigma(K_1)\,\sigma(K_2)\,\hat{f} = (K_1K_2f)\hat{}$$

which implies that $K_1 \circ K_2 = K_1K_2$ on $\mathcal{D}$. From these facts and the continuity of our operators in L^p, we readily obtain conclusion (a).

If $\sigma(K)(\zeta')$ never vanishes on the unit sphere, it follows by continuity that $\sigma(K)$ is bounded away from zero. Hence its reciprocal, $\sigma(K)^{-1}$, is again a symbol. So there exists a B_β singular operator H such that $\sigma(H) = \sigma(K)^{-1}$. Then, by formula (i), for all $f \in \mathcal{D}$,

$$(HKf)\hat{} = \sigma(H)\,\sigma(K)\,\hat{f} = \hat{f}$$

which implies that $HK = I$ on $\mathcal{D}$, and hence, by continuity, on any L^p, $1 < p < \infty$.

Q.E.D.

As we mentioned at the end of Chapter II (see Exercise 3)

the operator Λ, given by

$$\Lambda = \sum_{j=1}^{n} R_j D_j$$

is a continuous map of L^p_{k+1} into L^p_k, $1 < p < \infty$, which commutes with differentiations and with the Riesz transforms R_j. In the following theorem, we shall examine the relationship between K^* and $K^\#$, and between K_1K_2 and $K_1 \circ K_2$, in the general case, as well as the boundedness of the commutator $K\Lambda - \Lambda K$.

<u>THEOREM</u>. 4.9. Let us denote by $\mathcal{K}_\beta$ the class of all B_β singular integral operators K, with the norm $\|K\|_\beta$.

(a) $\mathcal{K}_\beta$ is a Banach space. If K_1 and K_2 belong to $\mathcal{K}_\beta$ then

$$\|K_1 + K_2\|_\beta \leq \|K_1\|_\beta + \|K_2\|_\beta \quad \text{and} \quad \|K_1 \circ K_2\|_\beta \leq C\|K_1\|_\beta\|K_2\|_\beta$$

where C depends only on β.

(b) If $\beta > 1$, the operators $K_1K_2 - K_1 \circ K_2$ and $K_1^* - K_1^\#$ map continuously L^p_k into L^p_{k+1} for every $1 < p < \infty$ and $-\beta \leq k \leq$ $\leq \beta - 1$, and for all $f \in L^p_k$

$$\|(K_1K_2 - K_1 \circ K_2) f\|_{p,k+1} \leq C\|K_1\|_\beta\|K_2\|_\beta\|f\|_{p,k}$$

$$\|(K_1^* - K_1^\#) f\|_{p,k+1} \leq C\|K_1\|_\beta\|f\|_{p,k}$$

where the constants C depend only on p and β.

(c) If $\beta > 1$, the operator $K\Lambda - \Lambda K$ is a continuous map of L^p_k into itself for every $1 < p < \infty$ and $|k| \leq \beta - 1$, and for all $f \in L^p_k$

$$\|(K\Lambda - \Lambda K)f\|_{p,k} \leq C\|K\|_\beta \|f\|_{p,k}$$

where C depends only on p and β.

<u>Proof</u>. (a) The completeness of $\mathcal{K}_\beta$ is a direct consequence of the completeness of B_β and Definition 4.4. The norm inequalities follow easily from the definitions and Theorem 4.1.

(b) According to Proposition 4.6, let us expand the operators K_1 and K_2 in the form

$$K_1 = \sum_{m\geq 0} a_{\ell m} G^{\ell m} \quad \text{and} \quad K_2 = \sum_{\mu \geq 0} b_{\lambda\mu} G^{\lambda\mu}$$

where, for simplicity, we have set $a = a_{oo}$, $b = b_{oo}$, $G^{oo} = I$ the identity operator, and we have omitted writing the summations over the indices $\ell = 1, 2, \ldots, d(m)$ and $\lambda = 1, 2, \ldots, d(\mu)$.

By formula (9') in the proof of Theorem 4.5, we have the bounds

$$(1)\qquad \begin{cases} \| a_{\ell m} \|_\beta \le C(m+1)^{-3n/2} \, \| K_1 \|_\beta \\ \\ \| b_{\lambda\mu} \|_\beta \le C(\mu+1)^{-3n/2} \, \| K_2 \|_\beta . \end{cases}$$

Furthermore, we have that

$$K_1 K_2 = \sum a_{\ell m} G^{\ell m} b_{\lambda\mu} G^{\lambda\mu}$$

whereas, expanding $\sigma(K_1)$ and $\sigma(K_2)$ in terms of spherical harmonics, it follows readily that

$$K_1 \circ K_2 = \sum a_{\ell m} b_{\lambda\mu} G^{\ell m} G^{\lambda\mu} .$$

Consequently,

$$(2)\qquad K_1 K_2 - K_1 \circ K_2 = \sum a_{\ell m} \, (G^{\ell m} b_{\lambda\mu} - b_{\lambda\mu} G^{\ell m}) \, G^{\lambda\mu}$$

and the crux of the proof consists in estimating the operator $G^{\ell m} b_{\lambda\mu} - b_{\lambda\mu} G^{\ell m}$ which is a commutator of the singular convolution operator $G^{\ell m}$ and multiplication by the function $b_{\lambda\mu}$ in the class B_β, $\beta > 1$. The desired estimate is that for any $f \in L^p_k$, where $1 < p < \infty$ and $-\beta \le k \le \beta - 1$,

$$(3)\qquad \| (G^{\ell m} b_{\lambda\mu} - b_{\lambda\mu} G^{\ell m}) \, f \|_{p,k+1} \le C \| b_{\lambda\mu} \|_\beta \; m^{n/2} \| f \|_{p,k}$$

with C depending only on p, β (and the dimension n).

Let us momentarily postpone the proof of formula (3) and proceed with the rest of the proof of part (b).

Since the operators $G^{\ell m}$ are continuous in L^p_k, we deduce from (2) and (3) that, with various constants C depending at most on p, β and the dimension n,

$$\|(K_1K_2 - K_1 \circ K_2) f\|_{p,k+1} \le \sum \|a_{\ell m}(G^{\ell m} b_{\lambda\mu} - b_{\lambda\mu} G^{\ell m}) G^{\lambda\mu} f\|_{p,k+1} \le$$

$$\le C \sum \|a_{\ell m}\|_\beta \|(G^{\ell m} b_{\lambda\mu} - b_{\lambda\mu} G^{\ell m}) G^{\lambda\mu} f\|_{p,k+1} \le$$

$$\le C \sum \|a_{\ell m}\|_\beta \|b_{\lambda\mu}\|_\beta \, m^{n/2} \|G^{\lambda\mu} f\|_{p,k} \le$$

$$\le C \left\{ \sum_{\substack{m\ge 0\\ \mu\ge 0}} \|a_{\ell m}\|_\beta \|b_{\lambda\mu}\|_\beta \, m^{n/2} \right\} \|f\|_{p,k}$$

and, by (1) and the bound $d(m) \le Cm^{(n-2)}$,

$$\left\{ \cdot\ \cdot\ \cdot \right\} \le C\|K_1\|_\beta \|K_2\|_\beta \left[\sum_{m\ge 0} d(m)(m+1)^{-n} \right] \times$$

$$\times \left[\sum_{\mu\ge 0} d(\mu)(\mu+1)^{-3n/2} \right] \le C\|K_1\|_\beta \|K_2\|_\beta \times$$

$$\times \left[\sum_{m=1}^{\infty} m^{-2} \right] \left[\sum_{\mu=1}^{\infty} \mu^{-n/2-2} \right]$$

where the numerical series clearly converge. Hence,

$$\|(K_1K_2 - K_1 \circ K_2) f\|_{p,k+1} \le C\|K_1\|_\beta \|K_2\|_\beta \|f\|_{p,k}.$$

The corresponding estimate for $K_1^* - K_1^{\#}$ is even simpler to obtain. Since, by Theorem 3.18, the complex conjugate $\bar{\gamma}_m = (-1)^m \gamma_m$, using Plancherel's formula we see that the adjoint of $G^{\ell m}$ is equal to $(-1)^m G^{\ell m}$. So, from the expansion of K_1, we deduce that

$$K_1^* = \sum_{m\geq 0} (-1)^m G^{\ell m} \bar{a}_{\ell m}$$

where $\bar{a}_{\ell m}$ is the complex conjugate of $a_{\ell m}$.

On the other hand, expanding $\sigma(K_1)$ in spherical harmonics, we obtain

$$K_1^{\#} = \sum_{m\geq 0} (-1)^m \bar{a}_{\ell m} G^{\ell m}.$$

Consequently, if $f \in L_k^p$, $1 < p < \infty$, $-\beta \leq k \leq \beta - 1$, $\beta > 1$, we obtain from formulas (3) and (1) that

$$\|(K_1^* - K_1^{\#}) f\|_{p,k+1} \leq \sum \|(G^{\ell m}\bar{a}_{\ell m} - \bar{a}_{\ell m}G^{\ell m}) f\|_{p,k+1} \leq$$

$$\leq C \sum \|a_{\ell m}\|_\beta \; m^{n/2} \; \|f\|_{p,k} \leq$$

$$\leq C\|K_1\|_\beta \|f\|_{p,k}$$

where C depends only on p, β and the dimension n.

Returning now to the proof of formula (3), let us set $G = G^{\ell m}$, $m \geq 1$, $a = a(x) \in B_\beta$, $\beta > 1$, and let $f \in \mathcal{D}$. Consider the operator

$$(aG - Ga)\frac{\partial f}{\partial x_i} = \lim_{\varepsilon\to 0}\int_{|x-y|>\varepsilon} Y(x - y)[a(x) - a(y)]\, f_i(y)\, dy$$

where we have set $Y(z) = |z|^{-n}\, Y_{\ell m}(z')$ and $f_i = \frac{\partial f}{\partial x_i}$. Letting $a_j = \frac{\partial a}{\partial x_j}$, we have

$$(4) \qquad a(x) - a(y) = \sum_{j=1}^{n} (x_j - y_j)\, a_j(x) + b(x,y)$$

and the estimates

$$(5) \qquad \begin{cases} |a_j(x) - a_j(y)| \leq \|a\|_\beta |x - y|^\alpha \\ |b(x,y)| \leq n\|a\|_\beta |x - y|^{1+\alpha} \end{cases}$$

where $0 < \alpha = \beta - 1$ if $\beta \leq 2$ and $\alpha = 1$ if $\beta > 2$. (In (4), we could assume for simplicity that $a(x)$ is differentiable in the usual sense and then remove this restriction in the end by a passage to the limit).

Integrating by parts, we see that

$$\int_{|x-y|>\varepsilon} Y(x - y)[a(x) - a(y)]\, f_i(y)\, dy =$$

$$= \int_{|x-y|>\varepsilon} Y(x - y)\, a_i(y)\, f(y)\, dy +$$

$$+ \int_{|x-y|=\varepsilon} Y(x - y)[a(x) - a(y)]\, f(y)\, \gamma_i \varepsilon^{n-1}\, d\sigma +$$

$$+ \int_{|x-y|>\varepsilon} Y_i(x - y)[a(x) - a(y)]\, f(y)\, dy =$$

$$= I + II + \left\{ \int_{\varepsilon<|x-y|<1} + \int_{|x-y|\geq 1} \right\}$$

$$(Y_i(x - y)[a(x) - a(y)]\, f(y)\, dy)$$

where, in the integral II, γ_i is the i-th direction cosine of the normal to the surface $|x - y| = \varepsilon$ and $\varepsilon^{n-1} d\sigma$ is the element of area on that surface. Also, substituting (4) in the next to the last integral, we obtain

$$(6) \quad \begin{cases} \displaystyle\int_{|x-y|>\varepsilon} Y(x - y)[a(x) - a(y)]\, f_i(y)\, dy = I + II + \\ \displaystyle + \int_{\varepsilon<|x-y|<1} Y_i(x - y) \sum_j (x_j - y_j)\, a_j(x)\, f(y)\, dy + \\ \displaystyle + \int_{\varepsilon<|x-y|<1} Y_i(x - y)\, b(x,y)\, f(y)\, dy + \\ \displaystyle + \int_{|x-y|\geq 1} Y_i(x - y)[a(x) - a(y)]\, f(y)\, dy = \\ = I + II + III + IV + V. \end{cases}$$

Let us note that, in III, the functions $z_j Y_i(z)$ are homogeneous of degree $-n$ and have mean value zero on the sphere $|z| = 1$ (otherwise, if $f(y) \not\equiv 0$ and we take $a_j(x) = \delta_{ij}$, the integral III would diverge as $\varepsilon \to 0$, while it is not difficult to see that all the other terms in (6) would converge). Moreover, by Theorem 3.12, we have the bounds

$$(7)\quad \begin{cases} |Y(x-y)| \le Cm^{(n-2)/2}\,|x-y|^{-n} \\ |Y_i(x-y)| \le Cm^{n/2}\,|x-y|^{-n-1}. \end{cases}$$

Since $I = \int_{|x-y|>\varepsilon} Y(x-y)\,a_i(y)\,f(y)\,dy$, using Theorem 2.1.4 with $q = 2$ we obtain

$$\|I\|_p \le A_{p,2}\left\{\int_\Sigma |Y_{\ell m}|^2\,d\sigma\right\}^{1/2}\|a_i f\|_p \le C\|a\|_\beta\|f\|_p$$

where C depends only on p and n. The same bound of course holds for the limit of I as $\varepsilon \to 0$. Similarly, writing III in the form $III = \int_{|x-y|>\varepsilon} - \int_{|x-y|\ge 1}$ and applying Theorem 2.1.4 to each term we obtain, on account of (7), that

$$\|III\|_p \le \|a\|_\beta\,A_{p,2}\left\{\int_{|z|=1} |z_j Y_i(z)|^2\,d\sigma\right\}^{1/2}\|f\|_p \le$$

$$\le C\|a\|_\beta\,m^{n/2}\|f\|_p$$

with C depending only on p and n.

Using (5) and (7), we see that

$$|IV| \le Cm^{n/2}\,\|a\|_\beta\int_{|x-y|<1}|x-y|^{-n+\alpha}\,|f(y)|\,dy$$

$$|V| \le Cm^{n/2}\|a\|_\beta \int_{|x-y|\ge 1} |x - y|^{-n-1} |f(y)| \, dy$$

with constants C depending only on n. Consequently, by Young's theorem on convolutions with integrable kernels, we obtain

$$\|IV\|_p \le Cm^{n/2} \|a\|_\beta \|f\|_p$$

$$\|V\|_p \le Cm^{n/2} \|a\|_\beta \|f\|_p$$

where the constants C depend at most on p, n and β(since α depends on β). Clearly, the same estimates hold for the limits of IV and V as $\varepsilon \to 0$.

Finally, let us consider

$$II = \int_{|x-y|=\varepsilon} Y(x - y)[a(x) - a(y)] f(y) \gamma_i \varepsilon^{n-1} \, d\sigma.$$

Since $|a(x) - a(y)| \le n|x - y| \|a\|_\beta$ and $|\gamma_i| \le 1$, we see by virtue of (7) that

$$|II| \le n\|a\|_\beta \, Cm^{(n-2)/2} \int_{|x-y|=\varepsilon} |f(y)| \, d\sigma$$

hence, with a new constant C depending only on n, it follows that

$$|II_o| = \left|\lim_{\varepsilon\to 0} II\right| \le Cm^{n/2} \|a\|_\beta |f(x)|.$$

Consequently,

$$\|II_o\|_p \leq Cm^{n/2}\|a\|_\beta\|f\|_p.$$

Therefore, combining our estimates, we deduce that for all $f \in \mathcal{D}$

$$(8) \qquad \left\|(aG - Ga)\frac{\partial f}{\partial x_i}\right\|_p \leq Cm^{n/2}\|a\|_\beta\|f\|_p$$

for some constant C depending only on p, β and the dimension n.

Now, $\|(Ga - aG)f\|_{p,1} = \|(aG - Ga)f\|_{p,1}$ is equivalent to $\|(aG - Ga)f\|_p + \sum_{i=1}^{n}\|\frac{\partial}{\partial x_i}(aG - Ga)f\|_p$. Clearly,

$$\|(aG - Ga)f\|_p \leq C\|a\|_\beta\|f\|_p.$$

Moreover, since $\frac{\partial}{\partial x_i}$ commutes with $G = G^{\ell m}$, we see from (8) that

$$\|\frac{\partial}{\partial x_i}(aG - Ga)f\|_p \leq \|(\frac{\partial a}{\partial x_i}G - G\frac{\partial a}{\partial x_i})f\|_p + \|(aG - Ga)\frac{\partial f}{\partial x_i}\|_p \leq$$

$$\leq C\|a\|_\beta\|f\|_p + Cm^{n/2}\|a\|_\beta\|f\|_p \leq$$

$$\leq Cm^{n/2}\|a\|_\beta\|f\|_p.$$

Thus, we conclude that for all f in $\mathcal{D}$

$$(9) \qquad \|(Ga - aG)\, f\|_{p,1} \le Cm^{n/2}\|a\|_\beta \|f\|_p$$

with C depending only on p, β and the dimension n. Furthermore, since $\mathcal{D}$ is dense in L^p, it is clear that (9) holds for any $f \in L^p$. Hence, estimate (3) has been proved in the case $k = 0$.

Let us now show by induction that, for $0 \le k \le \beta - 1$,

$$(10) \qquad \|(Ga - aG)\, f\|_{p,k+1} \le Cm^{n/2}\|a\|_\beta \|f\|_{p,k}.$$

Suppose that (10) holds for $0 \le k = r - 1$, where $r \le \beta - 1$, and let $f \in L^p_r$. Since $a \in B_\beta$ and $\beta \ge r + 1$ then, with $a_i = \frac{\partial a}{\partial x_i}$, we have that $\|a_i f\|_{p,r} \le C\|a\|_\beta \|f\|_{p,r}$. Consequently,

$$\|\frac{\partial}{\partial x_i}(Ga - aG)\, f\|_{p,r} \le \|(Ga_i - a_i G)\, f\|_{p,r} + \|(Ga - aG)\frac{\partial f}{\partial x_i}\|_{p,r} \le$$

$$\le C\|a\|_\beta \|f\|_{p,r} + Cm^{n/2}\|a\|_\beta \|\frac{\partial f}{\partial x_i}\|_{p,r-1} \le$$

$$\le Cm^{n/2}\|a\|_\beta \|f\|_{p,r}$$

from which it follows that $\|(Ga - aG)\, f\|_{p,r+1} \le Cm^{n/2}\|a\|_\beta \|f\|_{p,r}$.

It remains to show that (10) is valid also for $-\beta \le k \le 0$. We may assume that $f \in \mathcal{D}$ and let $q = p/(p - 1)$. If $0 \le k \le \beta$ then, taking the supremum over all $g \in \mathcal{D}$ with $\|g\|_{q,k-1} \le 1$

and using (10) in the case just proved, we obtain

$$\|(Ga - aG)\, f\|_{p,-k+1} = \|(aG - Ga)\, f\|_{p,-k+1} \leq$$

$$\leq C \sup |\langle (aG - Ga)\, f, g \rangle| (\|g\|_{q,k-1})^{-1} =$$

$$= C \sup |\langle f, (Ga - aG)\, g \rangle| (\|g\|_{q,k-1})^{-1} \leq$$

$$\leq C \|f\|_{p,-k} \|(Ga - aG)\, g\|_{q,k}\, (\|g\|_{q,k-1})^{-1} \leq$$

$$\leq Cm^{n/2} \|a\|_\beta \|f\|_{p,-k}.$$

This concludes the proof of estimate (3) and hence of part (b) of the theorem.

(c). Recalling that $\Lambda = \sum_{j=1}^{n} R_j D_j = \sum_{j=1}^{n} D_j R_j$ where R_j is a Riesz transform and $D_j = (2\pi i)^{-1} \frac{\partial}{\partial x_j}$, we begin with the following observation.

CLAIM: If $K \in \mathcal{K}_\beta$, $\beta \geq 1$, and $|k| \leq \beta - 1$, then, as an operator on L^p_k, the commutator $D_j K - K D_j$ belongs to $\mathcal{K}_{\beta-1}$ and has symbol given by $D_j \sigma(K)$.

In fact, according to 4.6, we can expand K in the series

$$K = \sum_{m \geq 0} a_{\ell m} G^{\ell m}$$

with $a_{\ell m}$ in B_β, and we recall from Theorem 4.5 that if f belongs to L^p_k, so does Kf. Since the $G^{\ell m}$ and D_j commute, we have

$$D_j Kf - KD_j f = \sum_{m\geq 0} [D_j(a_{\ell m} G^{\ell m} f) - a_{\ell m} D_j(G^{\ell m} f)]$$

and, using the identity $D_j(bg) = (D_j b)\, g + b(D_j g)$ with $b \in B_\beta$ and $g \in L^p_k$, we obtain

$$(D_j K - KD_j)\, f = \sum_{m\geq 0} (D_j a_{\ell m})\, G^{\ell m} f.$$

But, since $D_j a_{\ell m}$ are coefficients in the series expansion of a $B_{\beta-1}$ singular integral operator with symbol $D_j \sigma(K)$, the claim is established.

Consider the commutator

$$K\Lambda - \Lambda K = \sum_j (KD_j R_j - D_j R_j K) =$$

$$= \sum_j (KD_j - D_j K)\, R_j + \sum_j D_j(KR_j - R_j K).$$

If $f \in L^p_k$, $1 < p < \infty$ and $|k| \leq \beta - 1$, $\beta > 1$, then

$$(11) \qquad \|(K\Lambda - \Lambda K)\, f\|_{p,k} \leq \sum_j \|(KD_j - D_j K)\, R_j f\|_{p,k} +$$

$$+ \sum_j \|D_j(KR_j - R_j K)\, f\|_{p,k}.$$

By the preceding claim we have that

$$D_j K - KD_j = \sum_{m\geq 0} (D_j a_{\ell m})\ G^{\ell m}$$

and, since $\|G^{\ell m} R_j f\|_{p,k} \leq C\|f\|_{p,k}$ where C depends only on p and n, we deduce, using estimate (1), that

$$\begin{aligned}
\|(D_j K - KD_j)\ R_j f\|_{p,k} &\leq C\|f\|_{p,k} \sum_{m\geq 0} \|D_j a_{\ell m}\|_{\beta-1} \leq \\
&\leq C\|f\|_{p,k} \sum_{m\geq 0} \|a_{\ell m}\|_{\beta} \leq \\
&\leq C\|f\|_{p,k}\|K\|_{\beta} \sum_{m=1}^{\infty} d(m)\ m^{-3n/2} \leq \\
&\leq C\|K\|_{\beta}\|f\|_{p,k} \sum_{m=1}^{\infty} m^{-2-n/2} = \\
&= C\|K\|_{\beta}\|f\|_{p,k}.
\end{aligned}$$

On the other hand, for the second sum in (11), in view of the commutativity of the pseudo-product and the result obtained in part (b), we deduce that

$$\begin{aligned}
\|D_j(KR_j - R_j K)\ f\|_{p,k} &\leq \|(KR_j - R_j K)\ f\|_{p,k+1} \leq \\
&\leq \|(KR_j - K \circ R_j)\ f\|_{p,k+1} + \\
&+ \|(R_j \circ K - R_j K)\ f\|_{p,k+1} \leq C\|K\|_{\beta}\|f\|_{p,k}.
\end{aligned}$$

Therefore, combining the estimates, we conclude that

$$\|(K\Lambda - \Lambda K) f\|_{p,k} \le C\|K\|_{\beta}\|f\|_{p,k}.$$

Q.E.D.

The following simple consequence of the preceding theorem turns out to be very useful in many applications.

<u>COROLLARY</u>. 4.10. Let H and K belong to $\mathcal{K}_\beta$, $\beta > 1$. Then, for every $1 < p < \infty$ and $-\beta \le k \le \beta - 1$, the commutator $HK - KH$ maps L^p_k into L^p_{k+1} continuously, with norm $\le C\|H\|_\beta\|K\|_\beta$ where C is a constant depending only on p and β.

<u>Proof</u>. Since the pseudo-product is commutative,

$$HK - KH = (HK - H \circ K) + (K \circ H - KH)$$

so, applying part (b) of Theorem 4.9, we obtain the desired conclusion.

Q.E.D.

We also wish to point out the following result (see [2], Theorem 6) whose proof however will not be given here. Together with Theorem 4.12 below, it has served as a basis for many successful applications of the theory of singular integrals to the study of partial differential equations.

THEOREM. 4.11. Let $\mathcal{A}^p$ denote the algebra of bounded operators on L^p, $1 < p < \infty$, generated by all B_β singular integral operators, $\beta > 1$, and their adjoints. Then, the following conclusions hold:

a) There exists an algebra homomorphism σ_p of $\mathcal{A}^p$ onto the algebra of all functions $a(x,\zeta)$ which are homogeneous of degree zero in ζ, belong to C^∞ on $|\zeta| > 0$, and such that, for every $0 \leq |\alpha| \leq 2n$, the functions of x

$$\left(\frac{\partial}{\partial\zeta}\right)^\alpha a(x,\zeta)$$

are in B_β. For every B_β singular integral operator K, with symbol $\sigma(K)$, we have that $\sigma_p(K) = \sigma(K)$ and $\sigma_p(K^*) = \overline{\sigma(K)}$.

b) For any $H \in \mathcal{A}^p$, $\sigma_p(H) = 0$ if and only if there exists a constant $A > 0$, depending on H, such that $\|H\Lambda f\|_p \leq A\|f\|_p$ for every $f \in L^p_1$.

c) If $H \in \mathcal{A}^p$ and $\sigma_p(H)$ is bounded away from zero, then there exists an invertible operator $\tilde{H}$ in $\mathcal{A}^p$ such that $\sigma_p(\tilde{H}) = \sigma_p(H)$.

d) Every bounded operator on L^p which commutes with every operator in $\mathcal{A}^p$ is a multiple of the identity operator.

e) The algebras $\mathcal{A}^p$ and $\mathcal{A}^q$, corresponding to any two spaces

L^p and L^q, $1 < p,q < \infty$, are isomorphic and there exists a natural isomorphism ϕ of $\mathcal{A}^p$ onto $\mathcal{A}^q$ such that $\sigma_p = \sigma_q\phi$.

THEOREM. 4.12. Let $P(x,D)\,u = \sum_{|\alpha|=m} a_\alpha(x)\, D^\alpha u$ be a linear homogeneous differential operator of order m, with coefficients $a_\alpha(x)$ in B_β, $\beta \geq 0$. If $u \in L^p_m$, $1 < p < \infty$, then there exists a B_β singular integral operator K such that

$$P(x,D)\,u = K\Lambda^m u$$

with symbol, $\sigma(K) = P(x,\zeta)|\zeta|^{-m}$, where $P(x,\zeta) = \sum_{|\alpha|=m} a_\alpha(x)\zeta^\alpha$ is the characteristic polynomial of $P(x,D)$.

Proof. In exercise 3 at the end of Chapter II we encountered the formula $D_j = R_j\Lambda$, valid on L^p_1. Consequently, if $|\alpha| = m$ and $u \in L^p_m$, since the Riesz transforms R_j commute with Λ,

$$D^\alpha u = R_1^{\alpha_1} \cdot\cdot\cdot R_n^{\alpha_n}\, \Lambda^m u = R^\alpha \Lambda^m u$$

where $R = (R_1, \ldots, R_n)$ is the vectorial Riesz transform.

Letting $K = \sum_{|\alpha|=m} a_\alpha(x)\, R^\alpha$, we have that K is a B_β singular integral operator satisfying $P(x,D)\,u = K\Lambda^m u$. Since $\sigma(R_j) = \zeta_j/|\zeta|$, the symbol of K is given by

$$\sigma(K) = \sum_{|\alpha|=m} a_\alpha(x)\, \sigma(R^\alpha) = \sum_{|\alpha|=m} a_\alpha(x)\, \zeta^\alpha|\zeta|^{-m} = P(x,\zeta)|\zeta|^{-m}.$$

Q.E.D.

BIBLIOGRAPHY

[1] A. P. Calderón, "Integrales singulares y sus applicationes a ecuaciones differenciales hiperbolicas". Universidad de Buenos Aires, (1960).

[2] A. P. Calderón, A. Zygmund, "Singular integral operators and differential equations". Amer. J. Math., 79, (1957), 901-921.

[3] A. P. Calderón, "Lebesgue spaces of differentiable functions and distributions." Amer. Math. Soc., Proc. Symp. Pure Math., Vol. IV, (1961), 33-49.

[4] L. Hörmander, "Estimates for translation invariant operators in L^p spaces". Acta Math., 104, (1960), 93-139.

[5] J. Horváth, "Topological vector spaces and distributions", Vol. I. Addison-Wesley, (1966).

[6] U. Neri, "Singular integrals: an introduction". University of Maryland Lecture Notes No. 3, (1967).

[7] L. Schwartz, "Théorie des distributions", Tome I, II. Hermann, Paris, (1957, 1959).

Lecture Notes in Mathematics

Vol. 85: P. Cartier et D. Foata, Problèmes combinatoires de commutation et réarrangements. IV, 88 pages. 1969. DM 8,– / $ 2.20

Vol. 86: Category Theory, Homology Theory and their Applications I. Edited by P. Hilton. VI, 216 pages. 1969. DM 16,– / $ 4.40

Vol. 87: M. Tierney, Categorical Constructions in Stable Homotopy Theory. IV, 65 pages. 1969. DM 6,– / $ 1.70

Vol. 88: Séminaire de Probabilités III. IV, 229 pages. 1969. DM 18,– / $ 5.00

Vol. 89: Probability and Information Theory. Edited by M. Behara, K. Krickeberg and J. Wolfowitz. IV, 256 pages. 1969. DM 18,– / $ 5.00

Vol. 90: N. P. Bhatia and O. Hajek, Local Semi-Dynamical Systems. II, 157 pages. 1969. DM 14,– / $ 3.90

Vol. 91: N. N. Janenko, Die Zwischenschrittmethode zur Lösung mehrdimensionaler Probleme der mathematischen Physik. VIII, 194 Seiten. 1969. DM 16,80 / $ 4.70

Vol. 92: Category Theory, Homology Theory and their Applications II. Edited by P. Hilton. V, 308 pages. 1969. DM 20,– / $ 5.50

Vol. 93: K. R. Parthasarathy, Multipliers on Locally Compact Groups. III, 54 pages. 1969. DM 5,60 / $ 1.60

Vol. 94: M. Machover and J. Hirschfeld, Lectures on Non-Standard Analysis. VI, 79 pages. 1969. DM 6,– / $ 1.70

Vol. 95: A. S. Troelstra, Principles of Intuitionism. II, 111 pages. 1969. DM 10,– / $ 2.80

Vol. 96: H.-B. Brinkmann und D. Puppe, Abelsche und exakte Kategorien, Korrespondenzen. V, 141 Seiten. 1969. DM 10,– / $ 2.80

Vol. 97: S. O. Chase and M. E. Sweedler, Hopf Algebras and Galois theory. II, 133 pages. 1969. DM 10,– / $ 2.80

Vol. 98: M. Heins, Hardy Classes on Riemann Surfaces. III, 106 pages. 1969. DM 10,– / $ 2.80

Vol. 99: Category Theory, Homology Theory and their Applications III. Edited by P. Hilton. IV, 489 pages. 1969. DM 24,– / $ 6.60

Vol. 100: M. Artin and B. Mazur, Etale Homotopy. II, 196 Seiten. 1969. DM 12,– / $ 3.30

Vol. 101: G. P. Szegö et G. Treccani, Semigruppi di Trasformazioni Multivoche. VI, 177 pages. 1969. DM 14,– / $ 3.90

Vol. 102: F. Stummel, Rand- und Eigenwertaufgaben in Sobolewschen Räumen. VIII, 386 Seiten. 1969. DM 20,– / $ 5.50

Vol. 103: Lectures in Modern Analysis and Applications I. Edited by C. T. Taam. VII, 162 pages. 1969. DM 12,– / $ 3.30

Vol. 104: G. H. Pimbley, Jr., Eigenfunction Branches of Nonlinear Operators and their Bifurcations. II, 128 pages. 1969. DM 10,– / $ 2.80

Vol. 105: R. Larsen, The Multiplier Problem. VII, 284 pages. 1969. DM 18,– / $ 5.00

Vol. 106: Reports of the Midwest Category Seminar III. Edited by S. Mac Lane. III, 247 pages. 1969. DM 16,– / $ 4.40

Vol. 107: A. Peyerimhoff, Lectures on Summability. III, 111 pages. 1969. DM 8,– / $ 2.20

Vol. 108: Algebraic K-Theory and its Geometric Applications. Edited by R. M. F. Moss and C. B. Thomas. IV, 86 pages. 1969. DM 6,– / $ 1.70

Vol. 109: Conference on the Numerical Solution of Differential Equations. Edited by J. Ll. Morris. VI, 275 pages. 1969. DM 18,– / $ 5.00

Vol. 110: The Many Facets of Graph Theory. Edited by G. Chartrand and S. F. Kapoor. VIII, 290 pages. 1969. DM 18,– / $ 5.00

Vol. 111: K. H. Mayer, Relationen zwischen charakteristischen Zahlen. III, 99 Seiten. 1969. DM 8,– / $ 2.20

Vol. 112: Colloquium on Methods of Optimization. Edited by N. N. Moiseev. IV, 293 pages. 1970. DM 18,– / $ 5.00

Vol. 113: R. Wille, Kongruenzklassengeometrien. III, 99 Seiten. 1970. DM 8,– / $ 2.20

Vol. 114: H. Jacquet and R. P. Langlands, Automorphic Forms on GL (2). VII, 548 pages. 1970. DM 24,– / $ 6.60

Vol. 115: K. H. Roggenkamp and V. Huber-Dyson, Lattices over Orders I. XIX, 290 pages. 1970. DM 18,– / $ 5.00

Vol. 116: Séminaire Pierre Lelong (Analyse) Année 1969. IV, 195 pages. 1970. DM 14,– / $ 3.90

Vol. 117: Y. Meyer, Nombres de Pisot, Nombres de Salem et Analyse Harmonique. 63 pages. 1970. DM 6.– / $ 1.70

Vol. 118: Proceedings of the 15th Scandinavian Congress, Oslo 1968. Edited by K. E. Aubert and W. Ljunggren. IV, 162 pages. 1970. DM 12,– / $ 3.30

Vol. 119: M. Raynaud, Faisceaux amples sur les schémas en groupes et les espaces homogènes. III, 219 pages. 1970. DM 14,– / $ 3.90

Vol. 120: D. Siefkes, Büchi's Monadic Second Order Successor Arithmetic. XII, 130 Seiten. 1970. DM 12,– / $ 3.30

Vol. 121: H. S. Bear, Lectures on Gleason Parts. III, 47 pages. 1970. DM 6,– / $ 1.70

Vol. 122: H. Zieschang, E. Vogt und H.-D. Coldewey, Flächen und ebene diskontinuierliche Gruppen. VIII, 203 Seiten. 1970. DM 16,– / $ 4.40

Vol. 123: A. V. Jategaonkar, Left Principal Ideal Rings. VI, 145 pages. 1970. DM 12,– / $ 3.30

Vol. 124: Séminare de Probabilités IV. Edited by P. A. Meyer. IV, 282 pages. 1970. DM 20,– / $ 5.50

Vol. 125: Symposium on Automatic Demonstration. V, 310 pages. 1970. DM 20,– / $ 5.50

Vol. 126: P. Schapira, Théorie des Hyperfonctions. XI, 157 pages. 1970. DM 14,– / $ 3.90

Vol. 127: I. Stewart, Lie Algebras. IV, 97 pages. 1970. DM 10,– / $ 2.80

Vol. 128: M. Takesaki, Tomita's Theory of Modular Hilbert Algebras and its Applications. II, 123 pages. 1970. DM 10,– / $ 2.80

Vol. 129: K. H. Hofmann, The Duality of Compact Semigroups and C*- Bigebras. XII, 142 pages. 1970. DM 14,– / $ 3.90

Vol. 130: F. Lorenz, Quadratische Formen über Körpern. II, 77 Seiten. 1970. DM 8,– / $ 2.20

Vol. 131: A Borel et al., Seminar on Algebraic Groups and Related Finite Groups. VII, 321 pages. 1970. DM 22,– / $ 6.10

Vol. 132: Symposium on Optimization. III, 348 pages. 1970. DM 22,– / $ 6.10

Vol. 133: F. Topsøe, Topology and Measure. XIV, 79 pages. 1970. DM 8,– / $ 2.20

Vol. 134: L. Smith, Lectures on the Eilenberg-Moore Spectral Sequence. VII, 142 pages. 1970. DM 14,– / $ 3.90

Vol. 135: W. Stoll, Value Distribution of Holomorphic Maps into Compact Complex Manifolds. II, 267 pages. 1970. DM 18,– / $ 5.00

Vol. 136: M. Karoubi et al., Séminaire Heidelberg-Saarbrücken-Strasbourg sur la K-Théorie. IV, 264 pages. 1970. DM 18,– / $ 5.00

Vol. 137: Reports of the Midwest Category Seminar IV. Edited by S. MacLane. III, 139 pages. 1970. DM 12,– / $ 3.30

Vol. 138: D. Foata et M. Schützenberger, Théorie Géométrique des Polynômes Eulériens. V, 94 pages. 1970. DM 10,– / $ 2.80

Vol. 139: A. Badrikian, Séminaire sur les Fonctions Aléatoires Linéaires et les Mesures Cylindriques. VII, 221 pages. 1970. DM 18,– / $ 5.00

Vol. 140. Lectures in Modern Analysis and Applications II. Edited by C. T. Taam. VI, 119 pages. 1970. DM 10,– / $ 2.80

Vol. 141: G. Jameson, Ordered Linear Spaces. XV, 194 pages. 1970. DM 16,– / $ 4.40

Vol. 142: K. W. Roggenkamp, Lattices over Orders II. V, 388 pages. 1970. DM 22,– / $ 6.10

Vol. 143: K. W. Gruenberg, Cohomological Topics in Group Theory. XIV, 275 pages. 1970. DM 20,– / $ 5.50

Vol. 144: Seminar on Differential Equations and Dynamical Systems, II. Edited by J. A. Yorke. VIII, 268 pages. 1970. DM 20,– / $ 5.50

Vol. 145: E. J. Dubuc, Kan Extensions in Enriched Category Theory. XVI, 173 pages. 1970. DM 16,– / $ 4.40

Vol. 146: A. B. Altman and S. Kleiman, Introduction to Grothendieck Duality Theory. II, 192 pages. 1970. DM 18,– / $ 5.00

Vol. 147: D. E. Dobbs, Cech Cohomological Dimensions for Commutative Rings. VI, 176 pages. 1970. DM 16,– / $ 4.40

Vol. 148: R. Azencott, Espaces de Poisson des Groupes Localement Compacts. IX, 141 pages. 1970. DM 14,– / $ 3.90

Vol. 149: R. G. Swan and E. G. Evans, K-Theory of Finite Groups and Orders. IV, 237 pages. 1970. DM 20,– / $ 5.50

Vol. 150: Heyer, Dualität lokalkompakter Gruppen. XIII, 372 Seiten. 1970. DM 20,– / $ 5.50

Vol. 151: M. Demazure et A. Grothendieck, Schémas en Groupes I. (SGA 3). XV, 562 pages. 1970. DM 24,– / $ 6.60

Vol. 152: M. Demazure et A. Grothendieck, Schémas en Groupes II. (SGA 3). IX, 654 pages. 1970. DM 24,– / $ 6.60

Vol. 153: M. Demazure et A. Grothendieck, Schémas en Groupes III. (SGA 3). VIII, 529 pages. 1970. DM 24,– / $ 6.60

Vol. 154: A. Lascoux et M. Berger, Variétés Kähleriennes Compactes. VII, 83 pages. 1970. DM 8,– / $ 2.20